AF598021

Methods in Molecular Biology™

Series Editor
John M. Walker
School of Life Sciences
University of Hertfordshire
Hatfield, Hertfordshire, AL10 9AB, UK

For further volumes:
http://www.springer.com/series/7651

Neural Progenitor Cells

Methods and Protocols

Edited by

Brent A. Reynolds

Department of Neurosurgery, University of Florida, Gainesville, FL, USA

Loic P. Deleyrolle

Department of Neurosurgery, University of Florida, Gainesville, FL, USA

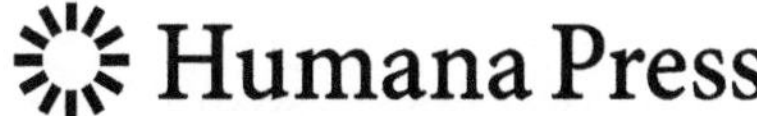

Editors
Brent A. Reynolds
Department of Neurosurgery
University of Florida
Gainesville, FL, USA

Loic P. Deleyrolle
Department of Neurosurgery
University of Florida
Gainesville, FL, USA

ISSN 1064-3745 ISSN 1940-6029 (electronic)
ISBN 978-1-62703-573-6 ISBN 978-1-62703-574-3 (eBook)
DOI 10.1007/978-1-62703-574-3
Springer New York Heidelberg Dordrecht London

Library of Congress Control Number: 2013943588

Printed on acid-free paper

Humana Press is a brand of Springer
Springer is part of Springer Science+Business Media (www.springer.com)

Preface

The discovery in the early 1990s of stem and progenitor cells in the adult mammalian CNS challenged the long-standing "no new neuron" doctrine attributed to the eminent scientist Ramon y Cajal and opened the door to the potential for cell replacement therapy. From discovery to clinical applications can be a long and tortuous route requiring rigorous steps involving standardized and precise protocols. This book is a collection of practical articles describing techniques used to study neural stem and progenitor cells and highlighting their promise toward stem cell-based therapeutic applications for CNS disorders.

We would like to express our appreciation to all authors for their valuable contribution to this volume, providing practical protocols that will assist the next generation of neuroscientists as they develop innovative experimental paradigms and progress towards therapeutic application.

Gainesville, FL, USA

Brent A. Reynolds
Loic P. Deleyrolle

Contents

Contributors

Hassan Azari • *Department of Anatomical Sciences, Neural Stem Cell and Regenerative Neuroscience Laboratory, Shiraz University of Medical Sciences, Shiraz, Iran*

Ricardo L. Azevedo-Pereira • *Department of Dermatology, Stanford University School of Medicine, California*

Perry F. Bartlett • *Queensland Brain Institute, The University of Queensland, Brisbane, QLD, Australia*

Luc Bauchet • *Département de Neurochirurgie, Centre Hospitalier Universitaire, Hôpital Gui de Chauliac, Montpellier, France; Institut des Neurosciences de Montpellier, INSERM U1051, Université Montpellier 2, Hôpital St Eloi, Montpellier, France; Centre Mutualiste Neurologique Propara, Montpellier, France*

Peter S. Bergin • *Faculty of Medical and Health Sciences, Centre for Brain Research, The University of Auckland, Auckland, New Zealand; Department of Neurology, Auckland City Hospital, Auckland, New Zealand*

Daniel G. Blackmore • *Queensland Brain Institute, University of Queensland, Brisbane, QLD, Australia*

Alain Boularan • *Département d'Anesthésie Réanimation, DAR C, Centre Hospitalier Universitaire, Hôpital Gui de Chauliac, Montpellier, France*

Chiara Cavazzin • *Division of Regenerative Medicine, Stem Cells and Gene Therapy, San Raffaele Scientific Institute, San Raffaele Telethon Institute for Gene Therapy (TIGET), Milan, Italy*

Andrew Chojnacki • *Department of Cell Biology and Anatomy, Faculty of Medicine, Hotchkiss Brain Institute, University of Calgary, Calgary, AB, Canada*

Maurice A. Curtis • *Department of Anatomy with Radiology, Faculty of Medical and Health Sciences, Centre for Brain Research, The University of Auckland, Auckland, New Zealand*

Marcel. M. Daadi • *Department of Neurosurgery, Stanford University School of Medicine, California & NeoNeuron, Palo Alto, California*

Mike Dragunow • *Department of Pharmacology, Faculty of Medical and Health Sciences, Centre for Brain Research, The University of Auckland, Auckland, New Zealand*

Richard L.M. Faull • *Department of Anatomy with Radiology, Faculty of Medical and Health Sciences, Centre for Brain Research, The University of Auckland, Auckland, New Zealand*

François Féron • *Aix Marseille University, NICN, CNRS UMR 7259, Center for Clinical Investigations in Biotherapy, Marseille, France*

Maurizio Gelati • *Laboratorio Cellule Staminali, Cell Factory e Biobanca Azienda Ospedaliera Santa Maria di Terni, Terni, Italy; IRCCS Casa Sollievo della Soffrenza, San Giovanni Rotondo, Foggia, Italy*

Stéphane D. Girard • *NICN, CNRS UMR 7259, Aix Marseille University, Marseille, France*

ANGELA GRITTI • *Division of Regenerative Medicine, Stem Cells and Gene Therapy, San Raffaele Scientific Institute, San Raffaele Telethon Institute for Gene Therapy (TIGET), Milan, Italy*

JEAN-PHILIPPE HUGNOT • *Hopital Saint Eloi, INSERM U1051, Institute for Neuroscience of Montpellier (INM), University Montpellier 2, Montpellier, France*

DHANISHA J. JHAVERI • *Queensland Brain Institute, The University of Queensland, Brisbane, QLD, Australia*

NICOLAS LONJON • *Département de Neurochirurgie, Centre Hospitalier Universitaire, Hôpital Gui de Chauliac, Montpellier, France*

SHARON A. LOUIS • *STEMCELL Technologies Inc., Vancouver, Canada*

ALAN MACKAY-SIM • *National Centre for Adult Stem Cell Research, Eskitis Institute for Drug Discovery, Griffith University, Brisbane, QLD, Australia*

CARMEN K.H. MAK • *STEMCELL Technologies Inc., Vancouver, Canada*

EDWARD W. MEE • *Faculty of Medical and Health Sciences, Centre for Brain Research, The University of Auckland, Auckland, New Zealand; Department of Neurosurgery, Auckland City Hospital, Auckland, New Zealand*

JOHANNA M. MONTGOMERY • *Department of Physiology, Faculty of Medical and Health Sciences, Centre for Brain Research, The University of Auckland, Auckland, New Zealand*

GIANMARCO MUZI • *Laboratorio Cellule Staminali, Cell Factory e Biobanca Azienda OSpedaliera Santa Maria di Terni, Terni, Italy*

MARGHERITA NERI • *Division of Regenerative Medicine, Stem Cells and Gene Therapy, San Raffaele Scientific Institute, San Raffaele Telethon Institute for Gene Therapy (TIGET), Milan, Italy*

GEOFFREY W. OSBORNE • *Queensland Brain Institute, The University of Queensland, St. Lucia, Australia; The Australian Institute for Bioengineering and Nanotechnology, The University of Queensland, St. Lucia, Australia*

CLARE L. PARISH • *Florey Neuroscience Institute, Melbourne Brain Centre, Parkville, VIC, Australia*

THOMAS I.H. PARK • *Department of Pharmacology, Faculty of Medical and Health Sciences, Centre for Brain Research, The University of Auckland, Auckland, New Zealand*

CHRIS PERRY • *National Centre for Adult Stem Cell Research, Eskitis Institute for Cell and Molecular Therapies, Griffith University, Brisbane, QLD, Australia; Department of Otolaryngology Head and Neck Surgery, Princess Alexandra Hospital, Brisbane, QLD, Australia*

STEVEN M. POLLARD • *Department of Cancer Biology, Samantha Dickson Brain Cancer Unit, UCL Cancer Institute, University College London, London, UK*

ALAIN PRIVAT • *Institut des Neurosciences de Montpellier, INSERM U1051, Université Montpellier 2, Hôpital St Eloi, BP 74103 80, Montpellier, France*

DANIELA PROFICO • *Laboratorio Cellule Staminali, Cell Factory e Biobanca Azienda Ospedaliera Santa Maria di Terni, Terni, Italy*

MASSIMO PROJETTI-PENSI • *Laboratorio Cellule Staminali, Cell Factory e Biobanca Azienda OSpedaliera Santa Maria di Terni, Terni, Italy*

BORIS W. PROSPER • *Queensland Brain Institute, The University of Queensland, Brisbane, QLD, Australia*

RODNEY L. RIETZE • *Neusentis (Pfizer Inc.), Cambridge, UK*

GIADA SGARAVIZZI • *Laboratorio Cellule Staminali, Cell Factory e Biobanca Azienda OSpedaliera Santa Maria di Terni, Terni, Italy*

FLORIAN A. SIEBZEHNRUBL • *Department of Neurosurgery, McKnight Brain Institute, University of Florida, Gainesville, FL, USA*
DENNIS A. STEINDLER • *Department of Neurosurgery, McKnight Brain Institute, University of Florida, Gainesville, FL, USA*
LACHLAN H. THOMPSON • *Florey Neuroscience Institute, Melbourne Brain Centre, Parkville, VIC, Australia*
FLORENCE VACHIERY-LAHAYE • *Coordination Hospitalière de Prélèvement et Agence de Biomédecine, Hôpital Gui de Chauliac, Centre Hospitalier Universitaire, Montpellier, France*
ANGELO LUIGI VESCOVI • *Laboratorio Cellule Staminali, Cell Factory e Biobanca Azienda Ospedaliera Santa Maria di Terni, Terni, Italy; Università Bicocca di Milano, Milan, Italy; IRCCS Casa Sollievo della Sofferenza, San Giovanni Rotondo, Italy*
HENRY J. WALDVOGEL • *Department of Anatomy with Radiology, Faculty of Medical and Health Sciences, Centre for Brain Research, The University of Auckland, Auckland, New Zealand*
MICHAEL D. WEISS • *Department of Pediatrics, McKnight Brain Institute, University of Florida, Gainesville, FL, USA*
SAMUEL WEISS • *Department of Cell Biology and Anatomy, Faculty of Medicine, Hotchkiss Brain Institute, University of Calgary, Calgary, AB, Canada*
TONG ZHENG • *Department of Neurosurgery, McKnight Brain Institute, University of Florida, Gainesville, FL, USA*

Part I

Culturing Neural Precursor Cells from Primary Tissue

Chapter 1

Culturing Fetal Precursor Cells Using Free Floating Serum-Free Conditions

Andrew Chojnacki and Samuel Weiss

Abstract

The propagation of neural precursors in culture is an essential tool for the study of the signaling matrix that regulates their proliferation, self-renewal, and generation of terminally differentiated progeny. Neural precursors can be expanded in vitro using both adherent and non-adherent culture protocols. The culture of fetal human neural precursors in the absence of serum as free-floating clusters of cells termed neurospheres is described here.

Key words Neurosphere, Neural stem cell, Cell culture, Self-renewal, Multipotent, Oligodendrocytes, Astrocytes, Neurons, Expansion, Differentiation

1 Introduction

The discovery of cells in the adult murine brain that proliferated in the presence of epidermal growth factor (EGF) to form clusters of floating spheres (now termed neurospheres) [1] contributed to dispelling the dogma that neural stem cells were absent in the adult mammalian central nervous system. This novel serum-free approach for isolating and expanding neural precursor cells as free-floating neurospheres has led to rapid advancements in our understanding of neural stem cell biology, and has also been adapted for the expansion of brain cancer cells [2] and various nonneural precursors, such as carotid body stem cells [3]. The neurosphere culture system can be utilized to uncover and understand the signaling pathways that regulate neural precursor proliferation, survival, self-renewal, differentiation, and phenotype specification. For example, the formation of a single neurosphere under sufficiently low cell culture density conditions and in the presence of a factor of interest would indicate that the factor being tested is a mitogen for the neural precursors being examined. In contrast, the regulation of neural precursor self-renewal can be studied by examining whether a single neurosphere exposed

Brent A. Reynolds and Loic P. Deleyrolle (eds.), *Neural Progenitor Cells: Methods and Protocols*, Methods in Molecular Biology, vol. 1059, DOI 10.1007/978-1-62703-574-3_1, © Springer Science+Business Media New York 2013

to a factor of interest during its formation generates a greater or lesser number of secondary neurospheres when dissociated into a single cell suspension compared to a control neurosphere of equal size. Additionally, the neurosphere culture system can be used to examine tissues for the presence of neural precursor populations or unmask precursor potential that is normally suppressed by the in vivo environment [4]. Here, we expand an EGF-responsive neural precursor population from fetal human brain tissue using the free-floating neurosphere culture system. We also illustrate how plating cell density is critical to ensuring that the neurospheres generated are products of proliferation and not cell clumping or the aggregation of debris.

Therefore, we have demonstrated how the free-floating neurosphere culture system can be used to expand a population of fetal human neural precursors. Once a neural precursor is successfully expanded as a neurosphere, the signals that regulate its function can be readily and easily assessed.

2 Materials

2.1 Cell Culture

1. B27 (Invitrogen).
2. Human recombinant epidermal growth factor (EGF; Peprotech).
3. 0.04 % Trypan blue (w/v).
4. 5,000 U Penicillin/5,000 μg Streptomycin.
5. Poly-L-ornithine: Reconstitute at 0.15 mg/ml in cell culture water. Autoclave to sterilize and aliquot in polypropylene conical tubes while still warm (*see* **Note 1**). Store at −20 °C for up to 6 months.
6. 10× Dulbecco's modified Eagle's media (DMEM)/F12 (1:1): Add DMEM to 125 ml of cell culture water (*see* **Note 2**). Next, add the F12, and mix until fully dissolved. Top up with cell culture water to 200 ml in a graduated cylinder. Sterilize using a bottle-top filter and store at 4 °C for up to 1 month (*see* **Note 3**).
7. Hormone mix: 1× DMEM/F12 (1:1), 230 μM insulin, 930 μM apo-transferrin, 185 nM progesterone, 560 μM putrescine, 210 nM sodium selenite, 0.56 % glucose (w/v), 0.10 % $NaHCO_3$, 4.6 mM HEPES. Store at −20 °C for up to 2 months. For 864 ml of hormone mix add the reagents in the following order: 300 ml tissue culture water, 80 ml 10× DMEM/F12 (1:1), 16 ml 30 % glucose (w/v), 12 ml 7.5 % $NaHCO_3$ (w/v), 4 ml 1 M HEPES, 352 ml tissue culture water, 800 mg apo-transferrin (*see* **Note 4**), 50 ml of 4 mg/ml insulin, 50 ml of 1.55 mg/ml putrescine, 80 μl 0.52 mg/ml

sodium selenite, and 80 µl 2 mM progesterone. After all the ingredients have dissolved, filter with a bottle-top filter and store at −20 °C for up to 2 months in appropriately sized aliquots.

8. 4 mg/ml insulin. Dissolve 200 mg of insulin with 4 ml of filtered 0.1 N HCl in a 50 ml falcon tube. Make up to 50 ml with cell culture water.
9. 2 mM progesterone. To a 1 mg vial of progesterone add 1.59 ml of 95 % ethanol (v/v). Store in a sterile tube at −20 °C for up to 6 months.
10. 1.55 mg/ml putrescine. In 50 ml of cell culture water dissolve 77.3 mg of putrescine. Prepare fresh.
11. 0.52 mg/ml sodium selenite. To a 1 mg vial add 1.93 ml of cell culture water. Can be stored for up to 6 months at −20 °C.
12. Media Hormone Mix (MHM): 1× DMEM:F12 (1:1), 0.66 % glucose (w/v), 2 mM glutamine, 14.6 mM $NaHCO_3$, 5 mM HEPES buffer, 23 µg/ml insulin, 93 µg/ml apo-transferrin, 19 nM progesterone, 56 µM putrescine, 21 nm sodium selenite. For 100 ml of MHM add the reagents in the following order through a bottle-top filter: 37.5 ml cell culture water, 2 ml 30 % glucose (w/v), 1.5 ml 7.5 % $NaHCO_3$ (w/v), 0.5 ml 1 M HEPES, 1 ml 200 mM glutamine, and 37.5 ml cell culture water. Add 10 ml of DMEM/F12 (1:1) and 10 ml of hormone mix directly to the bottle (these have been previously filtered) (*see* **Note 5**).
13. PBS glucose: 1× Dulbecco's PBS (calcium- and magnesium-free), 100 U/ml of penicillin, 100 µg/ml streptomycin, 0.6 % glucose. Through a bottle-top filter add the following components as listed: 75 ml cell culture water, 20 ml 10× Dulbecco's PBS (calcium- and magnesium-free), 4 ml penicillin/streptomycin, 4 ml 30 % glucose (w/v), and 97 ml cell culture water. Store at 4 °C for up to 1 week.

2.2 Cell Culture Equipment

1. 10 cm Petri dish.
2. 14 ml polypropylene round-bottom tubes.
3. 15 ml polypropylene conical tubes.
4. 24-well plates.
5. 40 µm cell strainer.
6. 50 ml polypropylene conical tubes.
7. Bottle-top filter (0.22 µm; low protein binding).
8. Circular coverslips #12, 10 oz.
9. Cotton-plugged glass Pasteur pipettes.
10. Glass bead sterilizer (Fine Science Tools; FST250).

11. Glass Pasteur pipettes.
12. Hemacytometer (Hausser Scientific, 0.1 mm deep).
13. Humidified tissue culture incubator (37 °C, 5 % CO_2).
14. Laminar flow hood.
15. Microdissecting instruments: 45° angled Dumont 5 forceps, 8 cm curved extra fine Moria spring scissors (Fine Science Tools).
16. Pyrex media bottles (100, 250, or 500 ml).
17. Tissue culture flasks: 25 cm^2, 75 cm^2, 175 cm^2 (Nunc).
18. Table-top centrifuge.
19. Water bath at 37 °C.

3 Methods

All procedures are performed at room temperature unless specified otherwise.

3.1 Generation of Primary Neurospheres

1. Use a 10 ml serological pipette to sort out pieces of fetal human brain tissue (fluffy and white in appearance) from your sample (*see* **Note 6**). Place the pieces into a 50 ml conical tube. Aspirate the majority of the liquid with a sterile Pasteur pipette being careful not to aspirate the tissue.
2. Wash the sample repeatedly with PBS glucose until all the red blood cells have been removed (*see* **Note 7**). The PBS glucose should now appear clear, and the white brain tissue should have settled to the bottom of the tube.
3. Aspirate the majority of the PBS glucose and transfer the brain tissue to a 10 cm Petri using a 10 ml serological pipette (*see* **Note 6**).
4. Using two sterilized 45° angled Dumont 5 forceps remove any meninges that remain attached to the brain tissue.
5. Using sterilized 45° angled Dumont 5 forceps and extra fine spring scissors cut the brain tissue into 1–5 mm^3 pieces.
6. Transfer in 10 ml of PBS glucose (*see* **Note 8**) into a 50 ml conical tube (*see* **Note 9**).
7. Insert a pre-wet, fire-polished, cotton-plugged Pasteur pipette (*see* **Note 10**) so that it is 1 cm from the bottom of the conical tube and gently triturate 40× (*see* **Note 11**).
8. Allow the large pieces to settle to the bottom for 5 min and transfer 5 ml of the cell suspension through a 40 μm cell strainer into a new 50 ml conical tube. Wash the cell strainer with 5 ml of PBS glucose.

9. Repeat **steps 7–8** until all the tissue is dissociated (*see* **Note 12**).
10. Wash the cell suspension 3× in 50 ml of PBS glucose, by spinning at 80 × *g* for 5 min.
11. After the last spin, aspirate the supernatant and resuspend in 15 ml of MHM. Triturate 15× with a 10 ml serological pipette to resuspend the cell pellet.
12. Mix 10 μl of the cell suspension with 10 μl of Trypan blue. Pipette 10 μl into a hemacytometer chamber. Under a 10× objective, count the number of phase-bright (viable) and blue cells (dead) in two of the four large corner squares. The cell counts are multiplied by 10,000 to obtain the number of cells per ml of suspension.
13. Plate the cell suspension at 20,000 live cells/ml (*see* **Note 13**) in the presence of the growth factor of your choice (EGF in this example). Use an appropriately sized tissue culture flask and place in an incubator set at 37 °C and 5 % CO_2.
14. Feed the flask every week by carefully removing half the medium and replacing with the same volume of medium containing double the concentration of EGF (or the growth factor of your choice). At a plating density of 20,000 cells/ml, primary fetal human brain tissue cultures generally require 2–3 weeks to form neurospheres (Fig. 2a) depending on the quality of the primary tissue.

3.2 Differentiation of Dissociated Primary Neurospheres

1. Transfer the contents of a 3-week-old flask into appropriately sized conical tubes and spin down the neurospheres at 80 × *g* for 5 min.
2. Remove the supernatant being careful not to disturb the cell pellet.
3. Resuspend in 4 ml of PBS glucose (*see* **Note 8**) in a 14 ml round-bottom tube, and triturate 50× using a pre-wet, fire-polished Pasteur pipette.
4. Allow undissociated neurospheres to settle to the bottom of the tube for 5 min.
5. Remove 3.5 ml of PBS glucose and transfer to a new 14 ml round-bottom tube.
6. Repeat **steps 3–6** until all the neurospheres have been dissociated.
7. Spin the dissociate down for 5 min at 80 × *g* and resuspend in 2 ml of MHM.
8. After performing a cell count as described in Subheading 3.1 **step 12**, plate the cells at 100,000 cells/ml on poly-L-ornithine-coated coverslips.

9. Allow the cells to differentiate for 1–7 days (*see* **Note 14**), fix with ice-cold 4 % paraformaldehyde for 20 min and assess phenotype using indirect immunocytochemistry.

3.3 Differentiation of Whole Neurospheres

1. Transfer the contents of a 3-week-old flask into a 50 ml conical tube, and allow all the neurospheres to settle to the bottom of the flask.
2. Without disturbing the neurospheres that have settled to the bottom of the tube, aspirate the medium.
3. Wash neurospheres by topping up the tube containing the neurospheres with 50 ml of medium and allowing them to settle to the bottom of the tube as in **step 1**. Aspirate medium and repeat this step (*see* **Note 15**).
4. Use a Gilson pipette to transfer some of the washed neurospheres onto a poly-L-ornithine-coated glass coverslips. Allow neurospheres to differentiate for 1–7 days in MHM or differentiation factors of interest (*see* **Note 14**).

3.4 Passaging and Self-Renewal Assays

1. Collect and dissociate the primary neurospheres as described in Subheading 3.2, **steps 1–7**.
2. To passage the cells, replate at 40,000 cells/ml in flasks containing 20 ml of MHM and 20 ng/ml of EGF. To perform self-renewal assays skip this step and proceed to **step 3**.
3. Plate the cells at 1,000 cell/ml in 96-well plates in MHM containing EGF. Feed every 7 days for a period of 2–3 weeks, by replacing half the medium with MHM containing twice the concentration of EGF.
4. After 2–3 weeks, assess the number of secondary neurospheres generated. The number of secondary neurospheres generated is a crude indicator of the self-renewal capacity of the primary neural precursors in the growth factor they were expanded in.
5. Alternatively, single neurospheres can be assessed for their capacity to generate secondary neurospheres. Collect and wash the neurospheres as describe in Subheading 3.3.
6. Using a 200 μl Gilson pipette, transfer individual and similarly sized neurospheres into the wells of a 96-well plate prefilled with MHM containing the growth factor of you choice (EGF in this example).
7. Triturate the neurospheres with a 200 μl Gilson pipette set to 180 while observing them with an inverted microscope. Use the minimum number of triturations to achieve a single cell dissociate.
8. Feed cultures every 7 days by removing half the MHM and replacing with MHM containing twice the concentration of EGF.
9. Count the number of neurospheres generated after 2–3 weeks in culture.

4 Notes

1. Aliquoting poly-L-ornithine while still warm prevents its loss from adhesion to the vessel it is reconstituted in.
2. Ensure DMEM is fully dissolved prior to the addition of F12. Addition of F12 to DMEM that is not fully in solution may result in the formation of a precipitate.
3. DMEM and F12 often contain phenol red as a pH indicator. Phenol red is a weak estrogen [5]. Use phenol red free alternatives if your neural precursor population may be responsive to estrogens.
4. Rinse vials with medium to ensure all the powder is dissolved.
5. B27, at a final concentration of 2 %, can also be added to supplement the MHM. Be aware that B27 contains retinyl acetate and triiodo-L-thyronine, which may inhibit neurosphere formation by some neural precursors. Alternatively, retinyl acetate-free B27 can be used. We have found that antioxidant-free B27 does not promote neurosphere formation, lending credence to the evidence that the redox state of a neural precursor may determine its differentiation status [6].
6. Widen the tip of the serological pipette by breaking off the tip while it is still within the sterile wrapper. Depending on the quality of the sample, this modification will allow large solid pieces to be transferred more easily to a sterile 50 ml conical tube without damaging the cells.
7. Take care to keep the brain tissue as intact as possible because it greatly eases removal of the red blood cells.
8. Magnesium- and calcium-free PBS glucose markedly decreases the number of triturations required to obtain a single cell suspension. However, it should be noted that Notch signaling is activated in the absence of extracellular calcium [7], which may impact neurosphere formation as Notch signaling is important in regulating precursor cell self-renewal [8]. MHM can be used as a substitute for PBS glucose, but with reduced cell viability.
9. The brain tissue should not take up more than 3 ml of the total volume. For larger samples divide in separate tubes as required.
10. The Pasteur pipette should be fire-polished to a diameter of approximately 450 μm (Fig. 1a) as too small an opening (Fig. 1b) will reduce cell viability, whereas too large an opening (Fig. 1c) will increase the number of triturations required to obtain a single-cell suspension and will also reduce cell viability.
11. Be careful not to introduce bubbles into the PBS glucose as this will reduce cell viability.

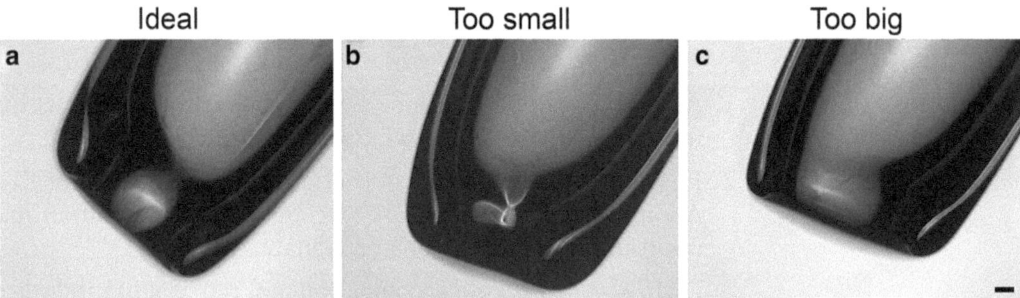

Fig. 1 Fire-polishing a Pasteur pipette to the ideal size for tissue and neurosphere dissociation. A Pasteur pipette that is 450 μm in diameter (**a**) typically results in the highest cell viability when used for mechanical dissociations of tissue or cultured neurospheres. Pipette tips that are fire-polished too much (**b**) or too little (**c**) can reduce cell viability by putting too much stress on the cells initially or by increasing the number of triturations required to attain a single cell suspension, respectively. Scale bars for **a** and **b** shown in **c**, 100 μm

12. We avoid enzymatically dissociating primary fetal human brain tissue as we find that it dissociates well mechanically. Enzymatic dissociation may also reduce the generation of neurospheres by reducing/altering the complement of proteins found on the cell membrane.
13. Primary cell plating density is absolutely critical to ensuring the neurospheres generated are products of mitogen-induced precursor cell proliferation and not cell clumping or debris aggregation. True neurospheres typically possess cilia-like projections that can be seen on the periphery of the sphere (arrows in Fig. 2b). In contrast, "neurospheres" composed of cell debris will lack these projections on their periphery (Fig. 2d). Dissociates contaminated with a large amount of debris and dead cells can generate neurosphere-like spheres at a density as low as 40,000 cell/ml even in the absence of growth factors (Fig. 2c, d). Cell density is especially critical in experiments attempting to demonstrate the presence of a precursor cell population in a particular tissue. For these types of experiments, we suggest only examining the flasks during feeding to prevent cell or debris aggregation caused by movement of the medium, or alternatively by plating the cells in 96-well plates. Furthermore, plating the cells at a clonal cell density (10,000 cells/ml or less) is absolutely required (*see* ref. [9]). It is important to note that high cell plating densities can inhibit neurosphere formation [9–11].
14. Incubation of the dissociated neurospheres in MHM is permissive for the differentiation of oligodendrocytes. Differentiation in 1 % fetal bovine serum tends to promote neuronal and astroglial differentiation at the expense of oligodendrocytes, which may be due in part to the presence of bone morphogenetic proteins in the serum [12].

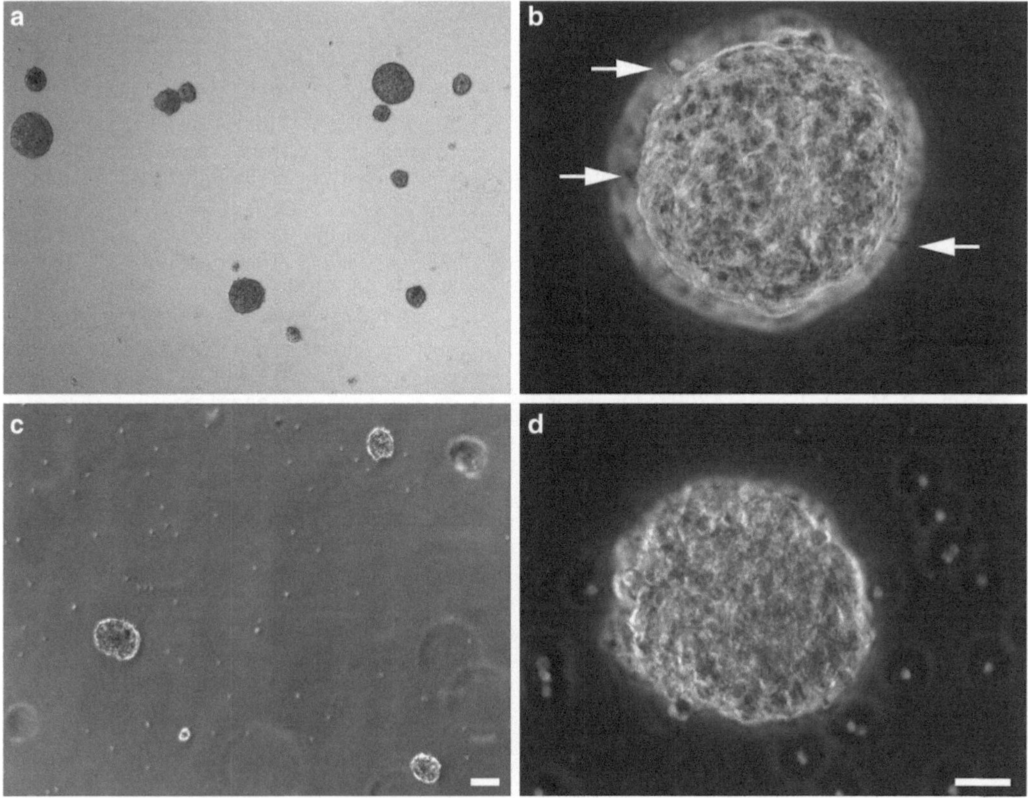

Fig. 2 Distinguishing true neurospheres from aggregates of debris and cells. (**a**) Two-week-old neurospheres generated in EGF at a plating density of 20,000 cells/ml. (**b**) Higher magnification of a neurosphere reveals multiple cilia-like processes protruding from the neurosphere periphery (*arrows*). Aggregates of debris, live and dead cells generated after 2 weeks in the absence of growth factors at a plating density of 40,000 cells/ml. (**d**) Higher magnification of the aggregate demonstrating a lack of cilia-like protrusions. Scale bar for **a**, shown in **c**, 100 μm; for **b**, shown in **d**, 50 μm

15. We have found (Chojnacki and Weiss, unpublished observations) that as little as 50 pg/ml of EGF is sufficient to promote proliferation of some fetal human neural precursors. Therefore, removal of growth factors in the whole neurosphere differentiation assay is particularly important to allow differentiation to proceed.

Acknowledgment

This work was supported by a Multiple Sclerosis Society of Canada Grant to A.C. and S.W., and a Canadian Institutes of Health Research grant to S.W.

References

1. Reynolds BA, Weiss S (1992) Generation of neurons and astrocytes from isolated cells of the adult mammalian central nervous system. Science 255(5052):1707–1710
2. Singh SK et al (2004) Identification of human brain tumour initiating cells. Nature 432(7015):396–401
3. Pardal R et al (2007) Glia-like stem cells sustain physiologic neurogenesis in the adult mammalian carotid body. Cell 131(2): 364–377
4. Buffo A, Rite I, Tripathi P, Lepier A, Colak D, Horn AP, Mori T, Gotz M (2008) Origin and progeny of reactive gliosis: a source of multipotent cells in the injured brain. Proc Natl Acad Sci USA 105(9):3581–3586
5. Berthois Y, Katzenellenbogen JA, Katzenellenbogen BS (1986) Phenol red in tissue culture media is a weak estrogen: implications concerning the study of estrogen-responsive cells in culture. Proc Natl Acad Sci USA 83(8):2496–2500
6. Noble M, Mayer-Proschel M, Proschel C (2005) Redox regulation of precursor cell function: insights and paradoxes. Antioxid Redox Signal 7(11–12):1456–1467
7. Rand MD et al (2000) Calcium depletion dissociates and activates heterodimeric notch receptors. Mol Cell Biol 20(5):1825–1835
8. Liu J, Sato C, Cerletti M, Wagers A (2010) Notch signaling in the regulation of stem cell self-renewal and differentiation. Curr Top Dev Biol 92:367–409
9. Coles-Takabe BL et al (2008) Don't look: growing clonal versus nonclonal neural stem cell colonies. Stem Cells 26(11):2938–2944
10. Chojnacki A et al (2008) Distinctions between fetal and adult human platelet-derived growth factor-responsive neural precursors. Ann Neurol 64(2):127–142
11. Capela A, Temple S (2002) LeX/ssea-1 is expressed by adult mouse CNS stem cells, identifying them as nonependymal. Neuron 35(5):865–875
12. Mabie PC et al (1997) Bone morphogenetic proteins induce astroglial differentiation of oligodendroglial-astroglial progenitor cells. J Neurosci 17(11):4112–4120

Chapter 2

In Vitro Expansion of Fetal Neural Progenitors as Adherent Cell Lines

Steven M. Pollard

Abstract

In vitro studies of neural progenitors isolated from the developing mouse have provided important insights into intrinsic and extrinsic pathways that control their behavior. However, use of primary cultures or neurospheres established from fetal tissues in cell population-based assays can be compromised by cellular heterogeneity. A complementary approach that addresses this issue is the establishment of adherent clonal neural stem (NS) cell lines. Here I describe protocols and troubleshooting advice for establishing adherent NS cell lines from the mouse fetal forebrain. NS cells grow as pure cultures in defined serum-free conditions as adherent monolayers and are therefore amenable to chemical/genetic screens, biochemical studies, and population-based analysis of gene expression or transcriptional regulation (e.g. RNA-Seq and ChIP-Seq). NS cell lines therefore represent a tractable cellular model system to explore the molecular and cellular biology of neural stem cell self-renewal and differentiation. Similar protocols can be extended to rat and human embryos, as well as human brain tumors.

Key words Mouse, Development, Forebrain, Neural stem cell, Adherent, Radial glia, Epidermal growth factor (EGF), Fibroblast growth factor, Laminin

1 Introduction

Mammalian stem cell lines, such as embryonic stem (ES) cells, are an important research tool that has been used to explore the molecular basis of pluripotency and regulation of early mammalian development. Establishing permanent cell lines as a model system is advantageous as a standardized resource is generated that can be shared between laboratories. The requirement of a constant supply of embryos is also avoided through continuous propagation of stem cells in vitro. As stem cell lines can be propagated indefinitely, they are well suited to those experimental approaches which require large numbers of cells, such as genome manipulations, high throughput or high content chemical/genetic screening, biochemical analysis, and gene expression profiling.

Brent A. Reynolds and Loic P. Deleyrolle (eds.), *Neural Progenitor Cells: Methods and Protocols*, Methods in Molecular Biology, vol. 1059, DOI 10.1007/978-1-62703-574-3_2, © Springer Science+Business Media New York 2013

The developing and adult mammalian nervous system contains a variety of spatially and temporally distinct neural progenitors capable of differentiation into neurons or glia. In vitro analysis of primary neural stem and progenitor cells using adherent colony forming assays [1, 2], or in suspension culture as "neurospheres" [3, 4], has provided a useful tool to explore mechanisms of self-renewal and differentiation in the developing and adult nervous system. However, to complement these studies researchers have not traditionally exploited the advantages of continuous propagation of neural stem cells as clonal cell lines, and instead typically expand primary cultures for only short periods [5].

In this chapter I describe protocols that can be used to establish and sustain clonal neural stem (NS) cell lines from mouse fetal forebrain. NS cell lines can be propagated continuously in adherent culture without the need for oncogene-based immortalization strategies [6]. NS cells have key molecular features of radial glia, long thought to provide solely an architectural role during forebrain development, but which are now known to self-renew and display neuronal and glial differentiation potential—defining properties of neural stem cells [reviewed in 7, 8].

Adherent monolayer culture has several experimental advantages that simplify analysis of stem cell behavior. Most notable is the lack of spontaneous cell death and differentiation that are a feature of neurospheres. This is analogous to ES cells, which when grown adherently are exposed to uniform culture environment and lack spontaneous differentiation, but when placed in suspension begin to generate embryoid bodies. The value of the increased homogeneity of NS cell cultures is exemplified by recent studies that have used NS cell lines to define transcription factor binding sites (e.g. REST and Mash1 [9, 10]), global mapping of epigenetic marks [11, 12], and also in studies of reprogramming to pluripotency [13–15]. Thus, NS cell lines represent a useful "discovery tool" to uncover gene function and biochemical mechanisms that control cell fate.

Despite these experimental advantages, there are important limitations that should be considered when modeling normal development and physiology. A particular concern for the use of in vitro expanded cultures is the potential loss of positional identity and consequently restrictions in the types of neurons and glia that can be produced [16]. For NS cells this is likely due to the effects of sustained exposure to FGF-2 in vitro which leads to upregulation of ventral markers such as Olig2 and Mash1 and restricts neuronal differentiation to GABAergic interneurons [17, 18]. Also, FGF-2 induces EGFR expression, which is typically found in later fetal stages, and suggests culture conditions also affect aspects temporal identity [19, 20]. To date clonal expansion of fully "plastic" naïve neural progenitors that can be steered in vitro to any desired neuronal or glial subtype has remained elusive, although several studies have moved us towards this important goal [21–23]. Finally, as is the case with any in vitro expansion, one also needs to be cautious of the inevitable selective pressures in culture that can lead to genetic abnormalities. Thus, while there are many experimental advantages in working with in vitro NS cell lines, caution should prevail when extrapolating findings to normal development and physiology.

2 Materials

2.1 Reagents

- Dulbecco's modified Eagle's Medium/nutrient mix F-12 (DMEM/Ham's F-12) media with l-Glutamine (*see* **Note 1**) (PAA; E15-813) (1×).
- Glucose solution (2 M) (Sigma, G8644-100ML).
- MEM nonessential amino acids (NEAA) (PAA; M11-003 100 ml) (100×).
- HEPES buffer solution (1 M) (PAA; S11-001, 100 ml) (200×).
- Penicillin/streptomycin (PAA; P11-010, 100 ml) (100×).
- N2 Supplement (PAA; F005-004, 5 ml) (Use at half recommended; 200×).
- B27 Supplement (Invitrogen 17504-044) (Use at half recommended; 100×).
- 2-mercaptoethanol, 50 mM (Invitrogen 31350-010 20 ml).
- Bovine Albumin fraction V (BSA) solution, 7.5 % (wt/vol) (Aliquot into 10 ml stocks) (Invitrogen 15260-037, 100 ml).
- Laminin (Sigma, L2020) 1 mg/ml stock in PBS (1,000×).
- 1× Accutase solution (PAA; L11-007, 100 ml).
- Phosphate-buffered saline (PBS) without Ca^{2+} and Mg^{2+} (PAA; H15-011).
- DMSO Hybri-Max (Sigma, D26-50).
- Accutase (Sigma, A6964) (1×).
- Murine epidermal growth factor (EGF) (PeproTech EC Ltd, 315-09) 100 μg/ml stock in PBS (10,000×).
- Human fibroblast growth factor 2 (FGF-2) (PeproTech EC Ltd, 100-18B) 50 μg/ml stock in PBS (5,000×).
- BMP-4 (PeproTech, AF-120-05ET-10) 10 μg/ml stock in PBS (1,000×).
- Tissue culture plastics and associated plasticware (*see* **Note 2**).

2.2 Preparation

For *NS cell media* prepare 10 × 500 ml bottles of NS cell media as follows.

1. Thaw all components and bring to room temperature.
2. Use fresh stocks of sterile solutions and prepare in a tissue culture hood.
3. To each 500 ml bottle of DMEM/F-12.
   - Add 7.25 ml of sterile glucose.
   - Add 5 ml of NEAA.
   - Add 5 ml of Pen/Strep.
   - Add 2.25 ml of HEPES.
   - Add 800 μl of BSA solution (75 mg/ml).
   - Add 0.5 ml 2-mercaptoethanol (handle with care).
   - Add 2.5 ml of N2 (*see* **Note 3**).
   - Add 5 ml of B27.
4. Mix gently avoiding inverting bottle and store at −20 °C (*see* **Note 4**).

"*Complete media*" (*CM*) is NS cell media supplemented with EGF and FGF-2 to final concentration of 10 ng/ml (*see* **Note 5**), as well as Laminin (1 μg/ml). We find that Laminin can be added directly to the culture media as a supplement where it is still effective at promoting attachment, avoiding the need for pre-coating of tissue culture plastic (*see* **Note 6**).

"*Wash media*" (*WM*) contains all components of the CM media, excluding N2 and B27 supplements, growth factors, and Laminin.

3 Methods

The protocols described are based on procedures using the E13.5 mouse telencephalon (*see* **Note 7**). However, similar cultures can be obtained from alternative developmental stages/regions or from germinal centers in the adult forebrain [6, 24] or following differentiation of mouse ES cells [6]. With some modifications they are also effective for other species (e.g., rat and human) [25, 26]. They are also applicable to certain types of human brain cancer [27].

3.1 Removal of the Mouse Fetal Brain

1. Prepare work area and dissection tools by wiping down with 70 % IMS or ethanol.
2. Using a pair of watchmakers forceps remove mouse embryos from the uterus and clean up with 2–3 washes in ice-cold PBS in a petri dish.

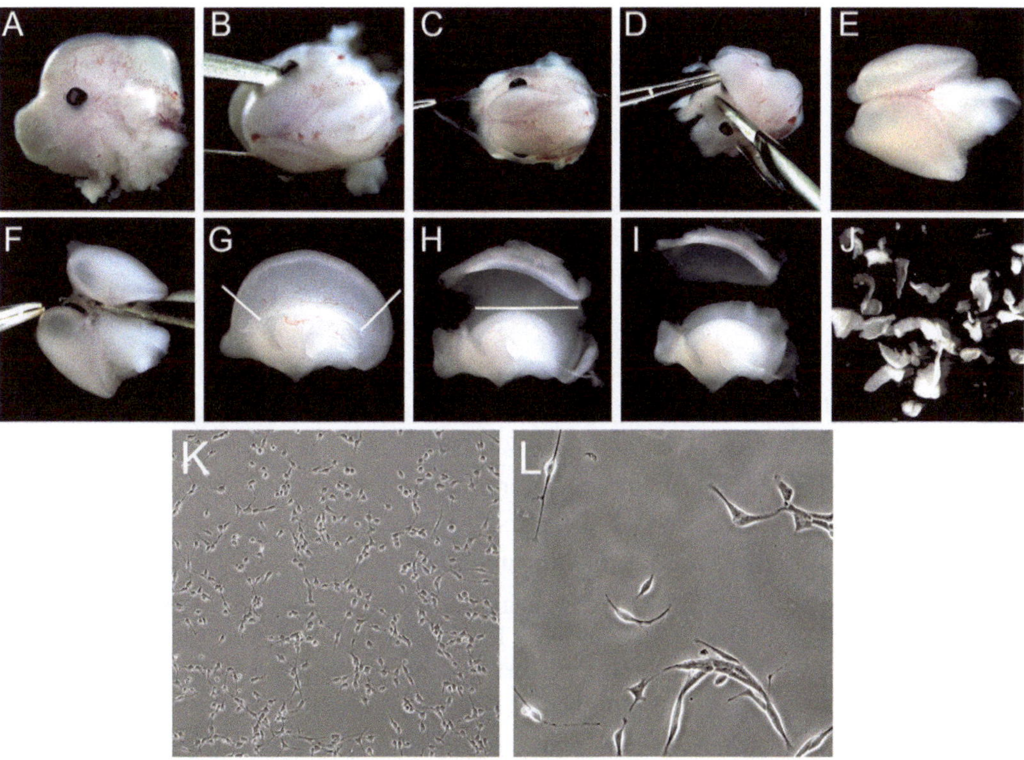

Fig. 1 Establishment of NS cell cultures from the E13.5 mouse embryo. (**a**)–(**e**) Isolation of the forebrain. (**f–j**) Dissection and dissociation of the dorsal forebrain. (**a**) Lateral view of the whole head. (**b**) Dorsal view of head anchored using forceps. (**c**) Removal of the skin. (**d**) Forebrain released following detachment of lateral facial tissues. (**e**) Isolated forebrain (telencephalic vesicles and diencephalon). (**f**) Separation of right telencephalic vesicle using forceps. (**g**) Medial surface of the telencephalic vesicle with remaining diencephalon tissues and ganglionic eminences visible. (**h**) Cortex released by making an incision at the rostral and caudal limits. (**i**) Detachment of the cortex by cutting above the dorsal edge of the LGE (*white line* in **h**). (**j**) Cortex is manually dissected into small pieces prior to dissociation in fresh PBS. (**k**) Appearance of NS cell cultures after several passages using phase-contrast microscopy (10×). (**l**) A clonal population of NS cell at higher power (20×). NS cell morphology is highly dynamic when monitored by time-lapse microscopy [6], but individual NS cells often display a typical bipolar radial glial-like morphology

3. Using forceps detach the head of each embryo and transfer to a fresh 3.5 cm petri dish filled with ice-cold PBS and place on ice (Fig. 1a).
4. Transfer one head to a fresh 10 cm petri dish filled with 10 ml of ice-cold PBS and place on dissecting microscope.
5. Anchor the head to the dish with dorsal surface facing upwards by pinning through the eye region with the forceps (Fig. 1b). Using the other forceps, pinch and then peel back the skin being careful not to damage the underlying tissues (Fig. 1c) (*see* **Note 8**).
6. Repeat this process to remove the dorsal portion of the developing skull.

7. Loosen the lateral developing facial tissues away from the brain by placing the forceps just below the eye and lifting/tearing caudally (Fig. 1d).
8. Separate the forebrain from the midbrain and gently lift it upwards to fully release (Fig. 1e). Discard the midbrain/hindbrain and remaining head tissues and transfer the forebrain into clean 10 cm dish containing 10 ml of ice-cold PBS.

3.2 Dissection of Dorsal Forebrain and Dissociation of Neural Progenitors

1. Release each of the telencephalic vesicles and discard the diencephalon (Fig. 1f).
2. Position each hemisphere with the medial surface facing upwards and make incisions near the rostral and caudal limits of the developing cerebral cortex (Fig. 1g).
3. Cut along the dorsal edge of the lateral ganglionic eminence to release the "flap" of cerebral cortex (Fig. 1h).
4. Remove the meningeal membranes (Fig. 1i) (*see* **Note 9**).
5. Transfer the isolated cortex to a fresh 3.5 cm dish (*see* **Note 10**), containing 1 ml of sterile PBS (Ca^{2+} and Mg^{2+} free) at room temperature and using the forceps chop the tissue into ~0.5 mm pieces (Fig. 1j).
6. Leave the tissue fragments for 5–10 min at RT to further dissociate and then triturate the undissociated tissue smoothly using a P1000 Gilson pipette for 1–2 min to achieve a single cell suspension (*see* **Note 11**).
7. Transfer the dissociated cell suspension to a 15 ml falcon tube containing 10 ml of pre-warmed wash media (WM) at 37 °C and mix by inversion.
8. To remove remaining non-dissociated tissue fragments pass the cell suspension through a cell strainer (75 μm) into a fresh 15 ml tube (*see* **Note 12**).
9. Centrifuge at room temperature for 5 min at 300 × *g* to obtain a pellet of dissociated cortical cells (*see* **Note 13**).

3.3 Establishing Adherent Neural Stem (NS) Cell Lines

1. Aspirate the supernatant and resuspend the dissociated cell pellet with 2 ml of complete media (CM) and transfer into one well of a 6-well plate. Incubate the plate at 37 °C, 5 % CO_2 (*see* **Note 14**).
2. Cells will expand rapidly as an adherent monolayer over the next few days (Fig. 1k). These primary cultures initially include some differentiated cells (*see* **Note 15**).
3. Once cells have reached ~80 % confluence (~1×10^5 cells) they can be transferred to a T25 flask as follows.
   (a) Remove the culture media and add 2 ml of pre-warmed PBS and remove immediately to wash away debris.

Replace this with 1 ml of pre-warmed Accutase solution (1×) and place in incubator for 2–5 min at 37 °C.

(b) The flask or plate should can be tapped against the bench to dislodge remaining cells, and then 5 ml of WM added. Smoothly pipette the cells up and down two to three times to fully dissociate.

(c) Transfer the dissociated cell suspension to a 15 ml falcon tube containing 5 ml of wash media and spin for 3–5 min at 300 × *g* at room temperature to obtain a pellet of dissociated NS cells.

(d) Resuspend the cell pellet in 5 ml of CM and transfer to a fresh T25. Cells should reach confluence in 2 days and can then be continuously expanded using the same procedure (**step 3** (**a**)–(**d**)) (*see* **Note 16**). Cultures should have typical NS cell appearance with individual cells displaying features of radial glia-like morphology (Fig. 1k, l).

3.4 Cryopreservation

If cells are not required immediately, then they can be cryopreserved as follows.

1. From a ~80 % confluent T25 flask (~1.5×10^6 cells) harvest a cell pellet as described above (Subheading 3.3, **step 3**). Resuspend the pellet in 1.5 ml of "freeze media" (CM plus DMSO freshly prepared to a final concentration of 10 %) and transfer 0.5 ml to each of three cryovials (~0.5×10^6 cells). Label and store at −80 °C for up to several months, or transfer to liquid nitrogen for long-term storage.
2. Cryovials can be stored at −80 °C for several months. To recover the cells a single cryovial is rapidly thawed in a water bath at 37 °C, cells are then transferred to 10 ml of pre-warmed WM, harvested by centrifugation and plated into a T25 flask containing 10 ml fresh pre-warmed complete media (CM). Following attachment (several hours or overnight) media is replaced with a fresh 5 ml CM is added to remove any residual DMSO.

3.5 Differentiation

1. For differentiation to astrocytes.

   (a) Seed 5×10^4 NS cells into a well of a four-well plate (Nunc) in CM (without EGF or FGF-2) and supplemented with either 1 % fetal calf serum (FCS) or 10 ng/ml BMP-4.

   (b) After 48 h >95 % of cells will have exited the cell cycle and acquire a characteristic stellate morphology.

2. For differentiation to neurons (*see* **Note 17**).

   (a) Seed 5×10^4 NS cells into each well of a four-well plate (Nunc) in CM without EGF, but plus FGF-2 at 10 ng/ml (i.e. EGF withdrawal).

(b) The following morning exchange the media fully with CM supplemented with FGF2 (5 ng/ml) + an additional 0.5× B27 supplement (Gibco).

(c) After 4–5 days exchange medium to NS basal media mixed with Neurobasal media (Invitrogen) (1:1) and supplement with 0.5× B27 (Gibco) but no growth factors. Cells with neuronal morphology should emerge over the next 3–7 days. Replace half media with fresh every 3–4 days.

(d) In order to maintain neurons for longer periods, from day 14 of differentiation switch media to Neurobasal media supplemented with B27 + BDNF (10 ng/ml; R and D systems), but without N2. Replace half of the media with fresh every 3–4 days. Such conditions should enable neuronal survival for a further 3 weeks [28].

3.6 Clonal Analysis

NS cell cultures can be used to establish clonal lines through plating at low density or single cell deposition in 96 well plates. Colony formation also serves as a useful assay for quantitative analysis of self-renewal.

1. Dispense 5,000 cells into a 10 cm dish in CM.
2. Around 10 days later around 200–300 discrete NS cell colonies will appear with around 1,000–2,000 cells/colony (*see* **Note 18**).
3. Colonies can be "picked" for further expansion as clonal cell lines as follows.

   (a) Remove the culture media from the 10 cm dish leaving only ~0.5 ml.

   (b) In the tissue culture hood hold up the plate up and tilt to 45° so that the residual media accumulates at the bottom. Rotate the plate such that the desired colony is free from media.

   (c) Load a P20 pipette with 20 μl of CM, use the tip to scratch in the region of the colony (*see* **Note 19**). Pipette the media up and down quickly once to dislodge and capture the colony.

   (d) Transfer these cells into one well of a 48-well plate and allow cells to expand until near confluent before passaging into a six-well plate and subsequently T25 flask.

3.7 Transfection (Nucleofection)

The nucleofection protocol described below can be used to achieve 50–70 % transient transfection of plasmid DNA using the Amaxa nucleofector™ (Lonza). It can also be adapted to generate stable transgenic clonal lines if used with linearized plasmid DNA and antibiotic selection. NS cells can also be genetically modified through use of viral vectors (retro- or lentiviral based).

1. Transfection requires 1–5 × 10^6 cells and 2 μg of plasmid DNA.

2. Add 0.5 ml of Amaxa supplement to 2.25 ml of Nucleofector solution V (as manufacturers instructions) and mix. This solution is now stable for 3 months at 4 °C.
3. Pre-warm Nucleofector solutions at room temperature.
4. Pre-warm NS media +10 % serum for cell recovery.
5. Prepare plates and media, label and place in incubator to warm.
6. Detach NS cells using protocol described above (Subheading 3.3, **step 3**) and centrifuge to harvest cells.
7. Resuspend the pellet in appropriate amount of nucleofection solution V (100 μl for each reaction, 1×10^6 cells). Do not leave longer than 15 min. Add 2 μg of DNA/reaction to this. Cells are now ready for nucleofection.
8. Transfer 100 μl cell solution to cuvette (*see* **Note 20**).
9. Choose program T-030 (T-020 will give better survival but less efficient).
10. Add cuvette to holder and press X to start.
11. Transfer cells gently into plates, transfer to the cell culture incubator and allow them to attach.
12. Once attached (usually ~2–3 h later), exchange media with fresh CM, washing twice to remove traces of serum.
13. Add drug selection after at least 24 h, or sort cells using flow cytometry (if a fluorescent reporter has been used).

4 Notes

1. For live fluorescence imaging DMEM:F12 should be replaced with Eagles modified essential media (EMEM, Sigma, M2645-10L) have less riboflavin, which significantly reduces background autofluoresence.
2. We find that tissue culture plastics supplied by Iwaki are optimal for attachment of NS cells.
3. Ensure that the supplement is fully in solution by pre-warming in water bath to 37 °C. Problems with a decrease in general viability of cells have been observed due to reduced insulin activity within the N2 supplement. N2 can also be "home made" as described previously [29].
4. *Optional*: to test for sterility, remove a 10 ml aliquot and mix with Tryptose solution 1:1 and place in incubator. This solution will turn cloudy within 1–2 days if contaminated with bacteria.
5. EGF and FGF-2 are stored in 50 μl aliquots at −20 °C (at 100 μg/ml and 50 μg/ml stock solutions in PBS, respectively) and are added freshly to the culture media. Thawed vials can be stored at 4 °C and should be used within 2–3 weeks.

6. This reduces the need for "pre-coating" of plates and reduces costs (coating is 10 μg/ml 2–3 h incubation; when used as a supplement 1 μg/ml is needed). For cell lines that are prone to aggregate increasing laminin in the culture media to 2 μg/ml can often promote monolayer expansion. We have found Laminin from different suppliers can vary dramatically in effectiveness. Laminin-1 as supplied by Sigma (L2020) has proven most consistent.
7. Although an E13.5 embryo is illustrated in this protocol, similar NS cells have been established from any stage between E12.5 and E18.5 using the same dissection/protocol. If working with a specific transgenic strain, then each embryo must be kept separate during the whole procedure. The remaining tissues after removal of the brain should be kept for genotyping. We normally process one litter at a time (10–15 embryos dissected within 1–2 h). We do not routinely use a laminar flow hood for the dissections.
8. The embryo can be anchored by placing each prong of the forceps into the eye and pinning down onto plastic and this makes it easier to make the initial pierce in the central region of the developing skull; after E15.5 the skull is more developed and care should be taken to avoid damaging the underlying brain.
9. The meningeal membranes are more difficult to detach at E12.5. It can help to leave the tissue in PBS for 5 min before removal.
10. For transferring tissue pieces use a P1000 and cut the tip with scissors to increase the diameter to ~2 mm.
11. Do this smoothly avoiding air bubbles as much as possible as this will result in reduced cell viability. Remove around 800 μl at a time and "blast" against the bottom of the dish to promote dissociation.
12. For establishing NS cell lines full dissociation of the tissue is not required. However, for detailed studies of the full repertoire of primary progenitors, then enzyme-based dissociation protocols should be used to maximise recovery of the full cellular population [30].
13. A more defined/compact cell pellet is obtained by spinning down cells in wash media than using PBS.
14. The same media and handling of cells can be applied to differentiating ES cells to generate NS cells [6, 29].
15. If there is extensive debris or the cell density is high then cells will initially form aggregates or neurospheres. These can be harvested by centrifugation at 100 × g for 30s and placed into a fresh well, where attachment should occur within 24 h

(pre-coating of plastic with the Laminin can be used; 10 μg/ml in PBS at 37 °C for 3 h).

16. We typically expand the cultures up to passage 3 or 4 before freezing down stocks. At this point >99 % of cells are immunopositive for nestin, RC2, Olig2, and Sox2 [6]. NS cells can then be continuously expanded. Mouse NS cell lines are effectively immortal, and have been continuously expanded for over 100 passages without crisis or changes in differentiation potential [6]. However, as with any in vitro culture with higher passage numbers there is a risk of accumulating genetic defects [31]. It is recommended that a large number of vials are stored at early passages and experiments past passage 20 cultures are avoided where possible.
17. Modifications to the neuronal differentiation protocol have been reported that enable more efficient generation of oligodendrocytes [32]. Differentiation works best using tissue culture plastic precoated with Laminin.
18. For true clonal analysis single cells should be deposited into individual wells using a cell sorter or limiting dilution and visual inspection of wells. Efficient colony formation depends on B27 supplement and is much less efficient using only N2. Precoating with Laminin can help restrict the colony size.
19. Areas with colonies can be highlighted with a marker pen on the bottom of the flask in advance to help picking.
20. Add solution gently and do not mix, just withdraw cells smoothly and place into a 1.5 ml microfuge tube for 5 min.

Acknowledgments

I thank all members of the Pollard laboratory and Dr. Stefano Bartesaghi for many helpful comments on the manuscript. Ursula Grazini helped in imaging embryos during the dissection procedure. S.M.P. is supported by grants from The Brain Tumour Charity and Cancer Research UK. He holds the Alex Bolt Research Fellowship.

References

1. Davis AA, Temple S (1994) A self-renewing multipotential stem cell in embryonic rat cerebral cortex. Nature 372:263–266
2. Johe KK et al (1996) Single factors direct the differentiation of stem cells from the fetal and adult central nervous system. Genes Dev 10: 3129–3140
3. Reynolds BA, Weiss S (1992) Generation of neurons and astrocytes from isolated cells of the adult mammalian central nervous system. Science 255:1707–1710
4. Reynolds BA, Tetzlaff W, Weiss S (1992) A multipotent EGF-responsive striatal embryonic progenitor cell produces neurons and astrocytes. J Neurosci 12:4565–4574
5. Pastrana E, Silva-Vargas V, Doetsch F (2011) Eyes wide open: a critical review of sphere-formation as an assay for stem cells. Cell 8:486–498

6. Conti L et al (2005) Niche-independent symmetrical self-renewal of a mammalian tissue stem cell. PLoS Biol 3:e283
7. Ever L, Gaiano N (2005) Radial "glial" progenitors: neurogenesis and signaling. Curr Opin Neurobiol 15:29–33
8. Pollard SM, Conti L (2007) Investigating radial glia in vitro. Prog Neurobiol 83:53–67
9. Johnson R et al (2008) REST regulates distinct transcriptional networks in embryonic and neural stem cells. PLoS Biol 6:e256
10. Castro DS et al (2011) A novel function of the proneural factor Ascl1 in progenitor proliferation identified by genome-wide characterization of its targets. Genes Dev 25:930–945
11. Bernstein BE et al (2006) A bivalent chromatin structure marks key developmental genes in embryonic stem cells. Cell 125:315–326
12. Meissner A et al (2008) Genome-scale DNA methylation maps of pluripotent and differentiated cells. Nature 454:766–770
13. Blelloch R et al (2006) Reprogramming efficiency following somatic cell nuclear transfer is influenced by the differentiation and methylation state of the donor nucleus. Stem Cells 24:2007–2013
14. Silva J et al (2009) Nanog is the gateway to the pluripotent ground state. Cell 138:722–737
15. Kim JB et al (2008) Pluripotent stem cells induced from adult neural stem cells by reprogramming with two factors. Nature 454:646–650
16. Conti L, Cattaneo E (2010) Neural stem cell systems: physiological players or in vitro entities? Nat Rev Neurosci 11:176–187
17. Hack MA et al (2004) Regionalization and fate specification in neurospheres: the role of Olig2 and Pax6. Mol Cell Neurosci 25:664–678
18. Gabay L, Lowell S et al (2003) Deregulation of dorsoventral patterning by FGF confers trilineage differentiation capacity on CNS stem cells in vitro. Neuron 40:485–499
19. Ciccolini F et al (2005) Hellwig, prospective isolation of late development multipotent precursors whose migration is promoted by EGFR. Dev Biol 284:112–125
20. Pollard SM et al (2008) Fibroblast growth factor induces a neural stem cell phenotype in foetal forebrain progenitors and during embryonic stem cell differentiation. Mol Cell Neurosci 38:393–403
21. Elkabetz Y et al (2008) Human ES cell-derived neural rosettes reveal a functionally distinct early neural stem cell stage. Genes Dev 22:152–165
22. Koch P et al (2009) A rosette-type, self-renewing human ES cell-derived neural stem cell with potential for in vitro instruction and synaptic integration. Proc Natl Acad Sci USA 106:3225–3230
23. Falk A et al (2012) Capture of neuroepithelial-like stem cells from pluripotent stem cells provides a versatile system for in vitro production of human neurons. PLoS One 7:e29597
24. Pollard SM et al (2006) Adherent neural stem (NS) cells from fetal and adult forebrain. Cereb Cortex 16(Suppl 1):i112–i120
25. Sun Y et al (2008) Long-term tripotent differentiation capacity of human neural stem (NS) cells in adherent culture. Mol Cell Neurosci 38:245–258
26. Sun Y et al (2011) Interplay between FGF2 and BMP controls the self-renewal, dormancy and differentiation of rat neural stem cells. J Cell Sci 124:1867–1877
27. Pollard SM et al (2009) Glioma stem cell lines expanded in adherent culture have tumor-specific phenotypes and are suitable for chemical and genetic screens. Cell Stem Cell 4: 568–580
28. Spiliotopoulos D et al (2009) An optimized experimental strategy for efficient conversion of embryonic stem (ES)-derived mouse neural stem (NS) cells into a nearly homogeneous mature neuronal population. Neurobiol Dis 34:320–331
29. Pollard SM, Benchoua A, Lowell S (2006) Neural stem cells, neurons, and glia. Methods Enzymol 418:151–169
30. Capela A, Temple S (2006) LeX is expressed by principle progenitor cells in the embryonic nervous system, is secreted into their environment and binds Wnt-1. Dev Biol 291:300–313
31. Diaferia GR et al (2011) Systematic chromosomal analysis of cultured mouse neural stem cell lines. Stem Cells Dev 20:1411–1423
32. Glaser T et al (2007) Tripotential differentiation of adherently expandable neural stem (NS) cells. PLoS One 2:e298

Chapter 3

Isolate and Culture Precursor Cells from the Adult Periventricular Area

Chiara Cavazzin, Margherita Neri, and Angela Gritti

Abstract

Due to the complexity of the NSC niche organization, the lack of specific NSC markers and the difficulty of long-term tracking these cells and their progeny in vivo the functional properties of the endogenous NSCs remain largely unexplored. These limitations have led to the development of methodologies to efficiently isolate, expand, and differentiate NSCs ex vivo. We describe here the peculiarities of the neurosphere assay (NSA) as a methodology that allows to efficiently isolate, expand, and differentiate somatic NSCs derived from the adult forebrain periventricular region while preserving proliferation, self-renewal, and multipotency, the main attributes that provide their functional identification.

Key words Neural stem cells, Subventricular zone, Murine models, Cell cultures, Cell proliferation, Self-renewal, Multipotency, Brain

1 Introduction

Neural stem cells (NSCs) appear early in development and remain active within the central nervous system (CNS) for the whole life duration of the organism. During this developmental process they assume different cellular morphologies and reside within changing microenvironments (niches) while retaining the basic properties of a stem cell: multipotentiality and the ability to self renew. The complexity of the NSC niche organization, the lack of specific NSC markers and the difficulty of long-term tracking these cells and their progeny has limited the in vivo approaches aimed to study the functional properties of the endogenous NSCs. These limitations have led to the development of methodologies to efficiently isolate, expand, and differentiate NSCs ex vivo while preserving proliferation, self-renewal, and multipotency, the main attributes that provide their functional identification.

Studies from the early 1990s showed that NSCs can be isolated from virtually all the embryonic CNS regions and from the adult periventricular regions as well as along the whole neuroaxis of the

Brent A. Reynolds and Loic P. Deleyrolle (eds.), *Neural Progenitor Cells: Methods and Protocols*, Methods in Molecular Biology, vol. 1059, DOI 10.1007/978-1-62703-574-3_3, © Springer Science+Business Media New York 2013

adult CNS, expanded, genetically manipulated, and differentiated in vitro [1–5].

This chapter refers to the isolation, and extensive propagation of somatic NSCs derived from the adult murine SVZ niche under fully chemically defined, serum-free conditions by means of the NeuroSphere Assay (NSA).

This protocol, based on the use of growth factors for isolation and expansion NSCs in floating "neurospheres," has been firstly described in the early 1990s [4, 6, 7]. Over the last two decades it has been improved and successfully used as a valuable tool for isolating, selecting, maintaining, and enriching the embryonic and adult CNS stem cell populations in vitro. In fact, while stem cells in culture are characterized by the capacity to proliferate and self-renew extensively, thus giving rise to long-term expanding stem cell lines, transit amplifying progenitors display limited proliferative capacity without self-renewal, and are lost throughout extensive subculturing [8].

This method focuses on these four cardinal culture conditions: (1) low cell density, (2) absence of serum, (3) addition of the appropriate growth factors (i.e., EGF and/or FGF2), (4) absence of a strong cell adhesion substrate.

Under these conditions, cells from freshly dissociated neural tissues attach loosely to the substrate and most of them die in 2–3 days. During this time, a very small fraction of cells (1–2 %) become hypertrophic and begin to proliferate while remaining attached to the plate. Then, the progeny of these proliferating cells adhere to each other and form spherical clones that eventually detach from the plate, float in suspension, and give rise to the so-called "neurospheres." Importantly, only a fraction of cells in the neurospheres (5–10 % for adult neurospheres, but these values can change depending both on the age and on the brain region considered) have the ability to give rise to secondary neurospheres, as assessed by clonogenic assays [9]. Indeed, a neurosphere is made up by low numbers of stem cells and high numbers of proliferating/differentiating progenitors, and even terminally differentiated neurons and glia, and can be envisioned as the in vitro counterpart of the in vivo neurogenic compartment [10].

Neurospheres can be subcultured by mechanical or enzymatic dissociation and by replating under the same in vitro conditions. As in the primary culture, at every subculturing passage, differentiating/differentiated cells rapidly die while the NSCs proliferate, giving rise to secondary spheres that can then be further subcultured. This procedure can be sequentially repeated several times and, since each stem cell gives rise to many stem cells by the time a sphere is formed, it results in the expansion of the NSC population in culture.

In primary cultures cell density plays a key role. If an excessively high cell density is used cells of different types including differentiated cells, rapidly aggregate to form neurosphere-like clusters even in absence of cell proliferation. Lineage restricted progenitors can survive within these clusters for significant time in culture and, if subculturing is carried out under the same conditions, they can survive for the first few passages, being eliminated only gradually.

Similarly, subculturing is one of the most critical steps in NSC culturing. One stem cell can give rise to several daughter stem cells as well as to differentiating progeny, during the formation of the spheres that eventually undergo subculturing. The number of stem cells found in a sphere varies from different regions, age, or species; generally is higher in early postnatal than in adult-derived spheres. This is the first parameter that, together with the length of the cell cycle, determines the rate of amplification and the growth curve of NSCs in culture. Yet, upon subculturing, not all stem cells survive and generate secondary spheres. Thus, the subculturing efficiency is an additional parameter that needs to be taken into consideration when expanding NSCs in culture.

When properly performed, the NSA allows discriminating between stem cells and transiently dividing non-stem progenitors. The correct identification of the nature of the sphere-forming cell is a relevant issue. Since committed progenitors are endowed with limited proliferative capacity and can produce up to tertiary neurospheres, the definition of a neural cell as a stem cell (or of a population of cells as containing a stem cell) should be applied only to a founder cell (or population of cells) that self-renew extensively and can be propagated in long-term cultures. In this view, the generation of clonal neurospheres by the NSA needs to be monitored, by both clonogenic and population analysis, to document stem cell activity, and to identify bona fide stem cells. These assays must be performed over an extended period of time, to be measured not as absolute time frame but as the number of subculturing steps that cells undergo in a given timeframe [11]. It has been suggested that at least five subculturing passages are required to rule out the contribution of committed progenitors to the maintenance of the cell population [11]. As proliferation and self-renewal are necessary but not sufficient to identify a population of cells as containing stem cells it is mandatory to evaluate the long-term capability of NSCs to generate a progeny (neurons, astrocytes, and oligodendrocytes) several orders of magnitude more numerous than the starting population. In this chapter we also describe a simple protocol to differentiate NSC cultures in mixed glia/neuronal cultures in rather stable proportions, by plating them on an adhesion substrate and removing growth factors from medium.

Once established and functionally characterized, stem cell lines can be effectively expanded to obtain a large number of cells, which can be cryopreserved. This allows for the establishment of a reservoir of cells at early subculture passages, which can be further expanded to create a master cell "bank" for future experiments. In this chapter we finally describe how to cryopreserve NSC lines without affecting the NSC functional properties.

Establishment of in vitro settings necessarily results in disruption of the three-dimensional tissue structure, loss of specific cell-to-cell contacts and modification of the extracellular environment and signalling. This might also lead to alteration of biological and molecular properties and acquirement of stem cell features by committed progenitors. Thus, although the versatility shown by NSC cultures in vitro can be envisaged as an advantage, extreme caution is necessary when considering the potential in vivo translation. In light of the complexity of the biological concerns governing stem cell maintenance and differentiation, significant progress will require a close coordination between in vivo and in vitro approaches. In this scenario, in vitro systems of NSCs shall allow a deep analysis at cellular level providing useful information to be further validated in vivo in order to identify relevance to normal physiology.

2 Materials

2.1 Dissection and Dissociation

2.1.1 Instruments

1. Scissors, large.
2. 2 Scissors, small, pointed.
3. Scissors, curved, fine.
4. Forceps, large.
5. Forceps, small.
6. Spatula, small.
7. Spoon, perforated, round.
8. Scalpel.
9. Petri dishes, 100-mm sterile plastic.
10. Cell culture cluster 6 well, sterile, plastic.
11. Tubes, 50 mL plastic, sterile.
12. Tubes, 15 mL plastic, sterile.
13. 2 Beaker lined with gauze and filled with 70 % ethanol.
14. Filters for sterilization, 0.22 μm.
15. Syringe 50 mL.
16. Pipetman, 1,000 μm, and appropriate tips.
17. Pipetman, 200 μm, and appropriate tips.
18. Trypan blue.
19. Burker counting chamber.
20. Dissecting microscope.

2.1.2 Reagents

1. DPBS for dissection components: Dulbecco's Phosphate-Buffered Saline w/o Ca^{2+} and Mg^{2+}, 30 % Glucose, 1,000 U/mL Penicillin–1,000 μg/mL Streptomycin solution.
2. Papain.

3. 0.1 % Deoxyribonuclease I from bovine pancreas (Dnase).
4. Earle's balanced salt solution (EBSS).
5. L-Cysteine.
6. Ethylenediamine tetraacetic acid (EDTA).
7. Avertin: 1.5 % w/v 2,2,2 Tribromethanol 99 %, 2.5 % v/v 2-methyl-2-buthanol 99+ %.
8. 70 % Ethanol.
9. Dulbecco's Modified Eagle's Medium (DMEM).

2.2 Subculturing and Differentiation

2.2.1 Instruments

1. Pipetman, 10 μL, and appropriate tips.
2. Pipetman, 200 μL, and appropriate tips.
3. Pipetman, 1,000 μL, and appropriate tips.
4. Tubes, 15 mL plastic, sterile.
5. Tubes, 50 mL plastic, sterile.
6. Trypan blue.
7. Burker counting chamber.
8. Cell culture cluster 24 or 48 well, sterile, plastic.
9. Coverslips, round-glass, 12 or 10 mm diameter.
10. Petri dish, glass.
11. Sterilized fine forceps.

2.2.2 Reagents

1. 70 % Ethanol.
2. Matrigel growth factor reduced.
3. Dulbecco's Modified Eagle's Medium (DMEM).
4. Trypan blue.
5. Hormone mix 10× components: pure water, DMEM/F12 10× solution (without $NaHCO_3$; 120 g of DMEM/F12 1:1 powder in 1L pure water), 30 % glucose solution, 7.5 % $NaHCO_3$ solution, 1 M HEPES solution, apo-transferrin, human recombinant insulin solution (10 mg/mL), putrescine, 3 mM NaSelenite in water, 2 mM progesterone in 95 % EtOH.
6. *Control Medium.* pure water, DMEM/F12 10× solution, 30 % glucose solution, 7.5 % $NaHCO_3$ solution, 1 M HEPES solution, 1,000 U/mL penicillin–1,000 μg/mL streptomycin solution, 200 mM glutamine, hormone mix 10×.
7. *Complete medium.* control medium, 500 μg/mL EGF (epidermal growth factor 2), 100 μg/mL FGF-2 (fibroblast growth factor), 0.2 % heparin solution (heparin sodium salt from porcine intestinal mucosa, Grade I-A, ≥ 180 USP units/mg, powder in pure water).

8. *Progenitor medium.* control medium, 100 μg/mL FGF-2, 0.2 % heparin solution.
9. *Differentiating medium.* control medium, heat inactivated fetal bovine serum.

2.3 Cryopreservation

2.3.1 Instruments

1. Freezing jar.
2. Pipetman, 1,000 μL and appropriate tips.
3. Cryovials, 2 mL.
4. Tubes, 15 mL plastic, sterile.

2.3.2 Reagents

1. Isopropanol.
2. Freezing medium: Complete culture medium, 10 % Dimethyl sulfoxide.

3 Methods

3.1 Medium Preparation

1. Prepare hormone mix 10× stock. Combine in a sterile glass beaker 330 mL pure water, 40 mL DMEM/F12 10×, 8 mL 30 % Glucose, 6 mL 7.5 % $NaHCO_3$, 2 mL 1 M HEPES solution.
2. Add 400 mg Apo-transferrin.
3. Add 10 ml human recombinant insulin (10 mg/mL).
4. Weight 38.6 mg Putrescine, dissolve in 40 mL water and add to the hormone mix solution.
5. Add 40 μL 2 mM Progesterone and 40 μL 3 mM Na Selenite and add to the hormone mix solution.
6. Mix well and filter-sterilize. Aliquot in 50 mL tubes and store at −20 °C.
7. In order to prepare 1 L of Control medium combine in a sterile glass beaker 740 mL pure water, 100 mL DMEM/F12 10×, 20 mL Glucose 30 %, 15 mL $NaHCO_3$ 7.5 %, 5 mL HEPES 1 M, 10 mL Glutamine 200 mM, 10 mL 1,000 U/mL Penicillin–1,000 μg/mL Streptomycin solution, 100 mL 10× hormone mix.
8. Mix well and filter-sterilize. Store at 4 °C.
9. In order to prepare 1 L of Complete medium add to 1 L of Control medium, 2 mL 0.2 % heparin solution, 40 μL 500 μg/mL EGF, 100 μL 100 μg/mL FGF-2.
10. Mix well and filter-sterilize. Store at 4 °C.
11. In order to prepare 1 L of Progenitor medium add to 1 L of Control medium, 2 mL 0.2 % heparin solution, and 100 μL 100 μg/mL FGF-2.

12. Mix well and filter-sterilize. Store at 4 °C.
13. In order to prepare 1 L of Differentiating medium add to 980 mL of Control medium, 20 mL of heat inactivated fetal bovine serum.
14. Mix well and filter-sterilize. Store at 4 °C.

3.2 Dissection and Dissociation

1. For 50 mL digestion solution, weight out 950 U Papain, and transfer it in a sterile 50 mL tube. Weight out 10 mg Cysteine, and 10 mg EDTA, and transfer them into another sterile 50 mL plastic tube. Keep the tubes at 4 °C until the end of the dissection procedure.
2. Prepare 1 L of DPBS for dissection combining 980 mL DPBS, 10 mL 30 % Glucose, 10 mL 1,000 U/mL Penicillin–1,000 µg/mL Streptomycin solution. Mix well and filter-sterilize. Store at 4 °C.
3. Prepare some sterile 100 mm plastic Petri dishes with cold DPBS to transfer, wash, and dissect tissues.
4. Select tools needed to remove brain (large scissors, small pointed scissors, large forceps, and a small spatula). Immerse in gauze-lined beaker filled with 70 % ethanol (*see* **Note 1**).
5. Select tools for tissue dissection (small forceps, small scissors pointed, curved fine scissors, scalpel). Immerse in 70 % ethanol in the second gauze-lined beaker.
6. Warm EBSS and DMEM to room temperature.
7. Just prior to beginning the dissection procedure, add 30 mL EBSS to the tube containing Cystein–EDTA and vortex until the solution is clear. Add an additional 20 mL EBSS to the tube containing Papain and vortex. Mix the two solutions, add 500 µL of DNase stock solution and filter with a 0.22 µm filter. Keep the solution at 4 °C (*see* **Notes 2–4**).
8. Anesthetize mice by intraperitoneal injection of avertin, or use other approved anesthetic, and sacrifice them by cervical dislocation. Preparation of NSCs from neonatal or early postnatal mice requires a slightly different procedure. Please refer to **Note 5**.
9. Using large scissors cut off the head just above the cervical spinal cord region. Rinse the head with 70 % ethanol.
10. Using small pointed scissors make a medial caudal-rostral cut, and remove the skin of the head.
11. Using the small scissors, make a longitudinal cut through the skull along the sagittal suture. Be careful not to damage the brain.
12. Using large forceps, grasp and peel the skull of the right hemisphere outward to expose the brain. Repeat for the left hemisphere.

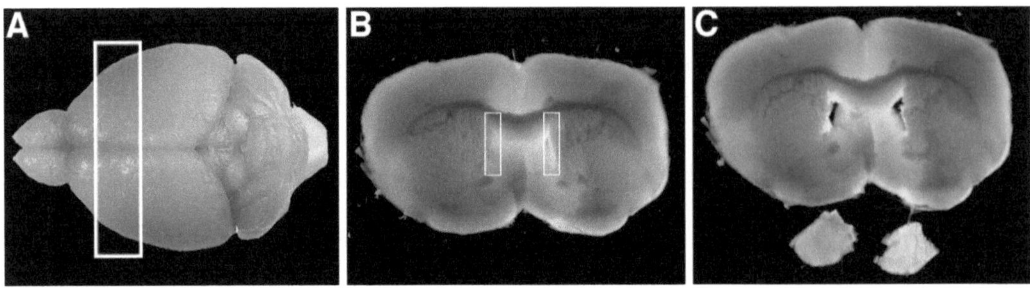

Fig. 1 Representation of tissue dissection. (**a**) Adult mouse brain. *Box* indicates the area to be cut in order to excide the SVZ. (**b**) Coronal section of the brain including the SVZ (*rectangles*). (**c**) Dissection of the SVZ

13. Using a small curved spatula, scoop the brain out and place in a Petri dish containing cold DPBS for dissection (*see* **Note 6**).
14. Wash brain twice by subsequently transferring them to new Petri dishes containing DPBS for dissection with the perforated round spoon.
15. To dissect the forebrain subventricular region, place the dish containing the brain under the dissecting microscope. Position the brain flat on its ventral surface, and hold it from the caudal side, using fine curved forceps.
16. Using the scalpel, make a coronal cut just behind the olfactory bulbs (Fig. 1a). Make a second coronal cut to dissect an approx 2-mm thick slice embodying the lateral ventricles (Fig. 1b). Discard the remaining pieces of the brain, and keep the slice.
17. Using fine forceps remove the residual choroid plexus inside the ventricles. Using fine curved microscissors, cut the thin layer of tissue surrounding the ventricles, excluding the striatal parenchyma and the corpus callosum (Fig. 1c). Place dissected tissue into a well of a six multiwell plate containing sterile DPBS for dissection.
18. Move the multiwell containing the tissues in the tissue culture laminar flow hood. From this point on, use aseptic technique.
19. Put 5 mL of Papain solution in each well of a new six multiwell and transfer there the dissected tissues using a P1000 pipette. Pool tissues from up to three mice in each well (*see* **Note 7**).
20. Incubate at 37 °C in the cells culture incubator for 30 min.
21. Transfer all the content of each well in a separate sterile 15 mL tube using a P1000 pipette and pipetting about five times disgregating the tissues, rinse the well with sterile DMEM, and add it to the tube. Pellet by centrifugation at $500 \times g$ for 10 min.
22. Remove almost all the supernatant (*see* **Note 8**).

23. Using a P1000 pipette add 500 μL of DMEM and dissociate 20–30× with the same tip. During trituration always avoid foaming and bubbles (*see* **Note 9**).
24. Add up to 12 mL of fresh DMEM to each tube and pellet the cells by centrifugation at 500 × *g* for 10 min.
25. Remove almost all the supernatant.
26. Using a P1000 pipette add 300 μL of DMEM and dissociate 20–30× with the same tip. Then use a P200 pipette with the volume set at 180 μL and dissociate 30–40× to obtain a single-cell suspension.
27. Add up to 12 mL of fresh DMEM to each tube and pellet the cells by centrifugation at 500 × *g* for 10 min.
28. Discard supernatant. Using a P200 pipette add 180 μL of the appropriate culture medium and dissociate 20–30× with the same tip.
29. Add medium up to 1 mL and dilute a 10 μL aliquot from each sample in trypan blue, and count in a Burker counting chamber (initially try a 1:2 dilution).
30. To initiate an NSC line seed cells at the appropriate density in complete culture medium, in untreated 6-multiwell tissue culture plates (5 mL volume), or 25 cm² tissue culture flasks (7 mL volume).
31. Incubate at 37 °C, 5 % CO_2, in a humidified incubator (*see* **Notes 10** and **11**).

3.3 Subculturing

1. Check cell cultures daily. After 2–3 days proliferating cells should start forming spherical clusters in suspensions, that we call neurospheres.
2. When the spheres reach about 100 μm diameter (5–10 days after plating, Fig. 2a) transfer content of the well to 15 mL sterile, plastic, conical tubes, using a sterile plastic pipette. Use 5 mL of fresh medium to rinse the well and add rinse to the tube.

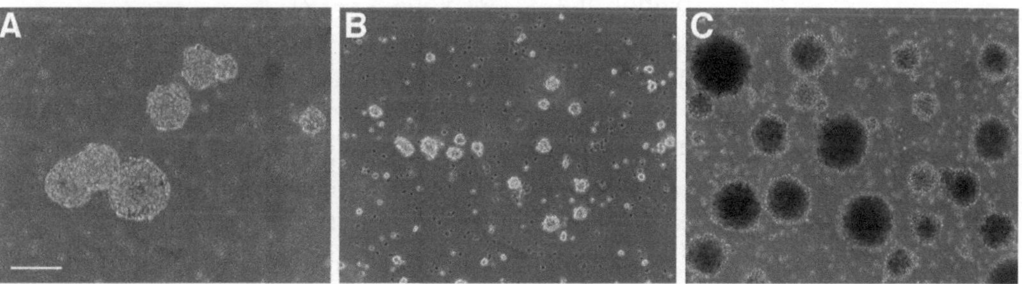

Fig. 2 Appearance of neurospheres in culture. (**a**) Neurospheres ready to be dissociated (approximately 100 μm diameter). (**b**) Small neurospheres. (**c**) Overgrown neurospheres. Scale bar: 100 μm

3. Pellet cell suspension by centrifugation at $110\times g$ for 10 min.
4. Remove the supernatant, leaving behind approx 50 μL. Using a sterilized p200 pipetman add 180 μL of fresh complete medium. Gently dissociate pellet 40–50×. Rinse down sides of tube at least after 20 pipetting, to dislodge undissociated spheres (*see* **Note 12**).
5. Count viable cells by trypan blue exclusion, and seed cells at 1×10^4 cells/cm^2 in fresh culture medium in untreated tissue flasks (*see* **Note 13**).
6. Subculture when the spheres growth up to 100 μm. This will require approx 5–7 days (*see* **Notes 14** and **15**).
7. Long-term proliferation of the bulk culture can be assessed by generating growth curves. Plate 2×10^5 viable cells in a 25 cm^2 flask (8,000 cell/cm^2) (0 days in vitro-DIV). At each subculture passage (every 5–7 days) count the total number of viable cells, and replate 2×10^5 cells under the same conditions. Repeat this subculturing procedure for at least 5–20 subculture passages.

   The estimated total number of cells is calculated by multiplying the amplification rate (total number of cells obtained at a given subculture passage/2×10^5) for the total number of cells obtained at the previous passage. Plot the total number of cells (in log10 scale) against the number of DIV. Interpolate data using a linear regression model and best fitted the following equation: $y=a+bx$, where y is the estimated total number of cells (in log scale), x is the time (DIV), a is the intercept, and b is the slope (*see* **Note 16**).

   Examples of ideal growth curves are shown in Fig. 3.

3.4 Assessment of Multipotency

1. The day before starting differentiation, immerse round-glass coverslips in 70 % ethanol. Dry with thin cotton gauze. Place in a glass Petri dish, and sterilize in preheated oven (at least 250 °C for 2 h).
2. The day before starting differentiation, prepare Matrigel diluting Matrigel stock solution 1:100 in sterile DMEM. Store the solution at 4 °C (*see* **Notes 17–19**).
3. The first day of differentiation, add one coverslip to each well of a 48- or 24-multiwell plate using sterilized fine forceps or a 2 mL plastic pipette connected to vacuum. Alternatively, glass or plastic chamber slides can be used.
4. Add 125 μL of Matrigel solution for each square centimeter of surface to each well. Incubate at 37 °C for at least 20 min.
5. Tap sides of flasks to dislodge spheres (*see* **Notes 20** and **21**).
6. Remove content of the flask to 15 mL sterile, plastic, conical tubes, using a sterile plastic pipet, and centrifuge at $110\times g$ for 10 min.

NSC #1				
subculturing passage	DIV	nr of cells plated	nr of cells obtained	total nr of cells
p5	0			2.00E+05
p6	5	2.00E+05	5.50E+05	5.50E+05
p7	10	2.00E+05	4.50E+05	1.24E+06
p8	15	2.00E+05	5.00E+05	3.09E+06
p9	20	2.00E+05	6.00E+05	9.28E+06
p10	25	2.00E+05	6.50E+05	3.02E+07
p11	30	2.00E+05	6.00E+05	9.05E+07

NSC #2				
subculturing passage	DIV	nr of cells plated	nr of cells obtained	total nr of cells
p5	0			2.00E+05
p6	5	2.00E+05	2.50E+05	2.50E+05
p7	10	2.00E+05	3.00E+05	3.75E+05
p8	15	2.00E+05	3.00E+05	5.63E+05
p9	20	2.00E+05	3.50E+05	9.84E+05
p10	25	2.00E+05	3.75E+05	1.85E+06
p11	30	2.00E+05	6.00E+05	5.54E+06

NSC #3				
subculturing passage	DIV	nr of cells plated	nr of cells obtained	total nr of cells
p5	0			2.00E+05
p6	5	2.00E+05	4.00E+05	4.00E+05
p7	10	2.00E+05	3.50E+05	7.00E+05
p8	15	2.00E+05	3.70E+05	1.30E+06
p9	20	2.00E+05	2.20E+05	1.42E+06
p10	25	2.00E+05	1.00E+05	7.12E+05
p11	30	2.00E+05	1.00E+05	3.56E+05

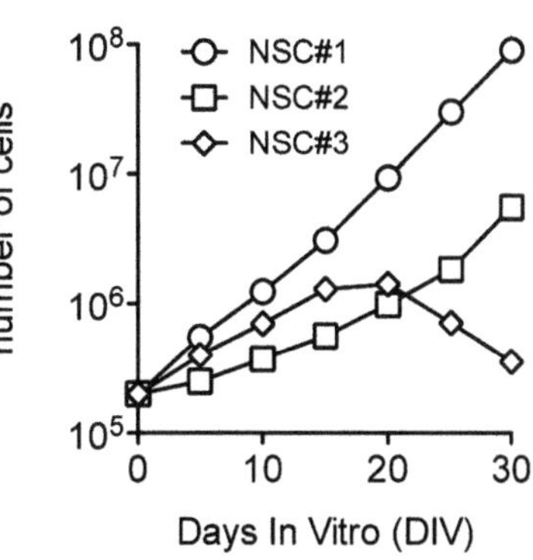

Fig. 3 Testing proliferation and long-term self-renewal of NSC cultures: growth rate calculation and plotting of three ideal NSC lines. NSC#1 represents a cell line with stable expansion rate, in which at each passage a similar number of cells (from two to threefold the number of plated cells) is retrieved. NSC#2 shows a slow rate of expansion at the first passages, then an increase in the growth rate. This situation is more frequent in the usual experimental settings. NSC#3 showing a initial phase of expansion and then a reduction of the number of cells retrieved at each passage, indicating the progressive loss of self-renewing precursors in the cell culture and the exhaustion of the cell line

7. To wash cells from growth factors, remove supernatant and resuspend cells gently, with a large-bore plastic pipet, in 10 mL control medium. Spin at 110 × *g* for 10 min.
8. Remove the supernatant, leaving behind about 200 μL. Using a sterilized p200 pipetman set at 180 μL, gently triturate pellet 50–60× (*see* **Note 22**).
9. Add Progenitor medium up to 1 mL and count viable cells by trypan blue exclusion. Resuspend cells in the appropriate volume of Progenitor medium, so that the final number of cells to be plated per well is: 35,000 cells in 0.5 mL for 48-multiwell plates or 70,000 cells in 1 mL for 24 multiwell plates.
10. Remove matrigel from coverslips (*see* **Note 23**).
11. Add appropriate volume of cell suspension to multiwells. Agitate plate to distribute cells evenly on coverslips (*see* **Note 24**).
12. Incubate at 37 °C in a humidified atmosphere of 5 % CO/95 % air for 72 h.
13. Remove carefully Progenitor medium and add 0.5 or 1 mL of differentiating medium to each well.
14. Stop the differentiation after 4–7 days. Representative bright field images of each step of differentiation is shown in Fig. 4.

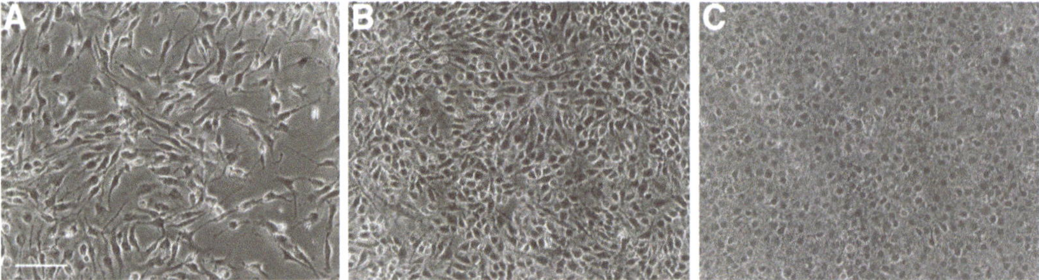

Fig. 4 Appearance of NSCs plated on Matrigel during the differentiation protocol. (**a**) 24 h in complete medium; (**b**) 72 h in Progenitor medium; (**c**) 5 days in Differentiating medium. Scale bar: 80 μm

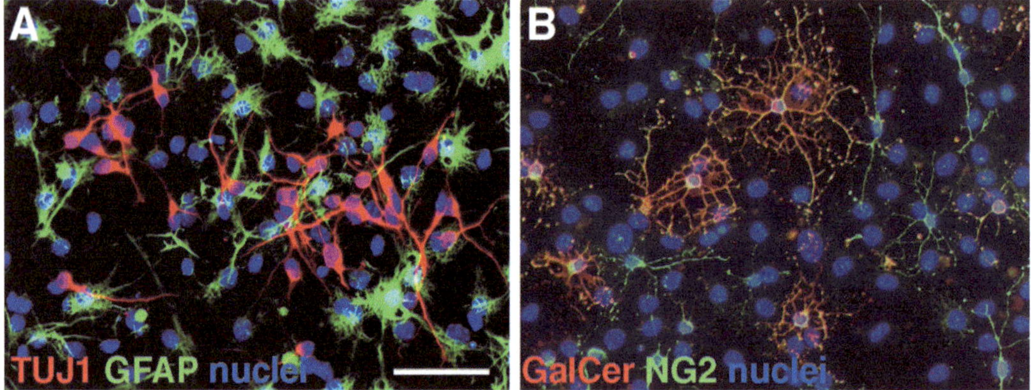

Fig. 5 Immunofluorescence analysis of differentiated NSC cultures. (**a**) Neuronal cells labeled with anti-beta tubulin class III (clone TuJ1) antibody (*red*) and astrocytes labeled with anti-glial fibrillar acidic protein (GFAP) antibody (*green*); (**b**) Oligodendrocytes labeled with anti-Galactocerebroside (GalC) antibody (*red*) and anti-proteoglycan NG2 (*green*). Nuclei stained with DAPI (*blue*). Scale bar: 100 μm

15. Wash the wells with DPBS then fix cells with 4 % paraformaldehyde solution, 10 min at room temperature (*see* **Notes 25–27**).
16. Proceed with the appropriate analysis (IF, IHC). An example of immunofluorescence for neurons, astrocytes, and oligodendrocytes is shown in Fig. 5.

3.5 Cryopreservation of Neurospheres

3.5.1 Freezing

1. Ensure that freezing jar is at room temperature and filled with isopropanol (*see* **Note 28**).
2. Prepare freezing medium and filter 0.22 μm. Consider 1.5 mL of freezing solution for each flask (*see* **Note 29**).
3. Label cryovials with date, cell type, and passage number and any other relevant additional information.
4. Collect spheres by gentle pipetting, and pellet them by centrifugation at 110×*g* for 10 min.
5. Remove the supernatant and resuspend neurospheres in 1.5 mL freezing medium gently pipetting 2–3× (*see* **Note 30**).

6. Transfer cells into labeled 2 mL cryogenic vials.
7. Transfer vials into the freezing jar containing isopropanol.
8. Leave the jar at −80 °C for a minimum of 4 h, to allow a slow and reproducible decrease in temperature (−1 °C/min).
9. Transfer vials into a liquid nitrogen tank for long-term storage.

3.5.2 Thawing of Cryopreserved Neurospheres

1. Warm culture medium in a water bath to 37 °C.
2. Quickly transfer cryovial from liquid nitrogen to 37 °C water bath, and leave until it is almost completely thawed (few minutes). Swirl the vial to favor thawing.
3. Wipe entire cryovial with 70 % ethanol.
4. Slowly transfer cell suspension from cryovial to 15 mL plastic tube containing 5 mL warm culture medium using a large-bore plastic pipette.
5. Spin cell suspension for 10 min at 110 × *g* and remove most of the supernatant.
6. Gently resuspend pellet in fresh medium (do not dissociate) and plate in flasks of appropriate size. Wait 24–48 hours before dissociating thawed neurospheres.

4 Notes

1. Use a gauze-lined beaker to avoid spoiling the tips of the microforceps and scissors.
2. Papain solution does not become clear until it is mixed with Cystein and EDTA solution; heating papain solution at 37 °C for 10 min can help solubilization.
3. Thirty minutes is the average time required for an experienced person to dissect tissues from two mice.
4. Euthanasia, removal and dissection of brain can be performed outside the laminar flow hood. Particular caution should be exercised to avoid contamination. Wear gloves and frequently rinse hands with ethanol.
5. In order to generate NSC lines from neonatal (PND0–PND2) and early postnatal mice (PND3–PND10) some variations to the described protocol are necessary.
   - The SVZ area in neonatal and early postnatal mice is larger than in adult mice and the tissue is less compact.
   - Skip papain digestion (**steps 18–20**) and proceed directly to mechanical dissociation (**step 21**) for tissues up to PND2. For tissues from mice older than PND2 apply a 10–15 min papain digestion step.
   - Being the neonatal SVZ richer in stem/progenitor cells than the adult, the cell yield is higher (about 2×10^5 from

each SVZ). Moreover, less dead cells and debris are present in the primary culture due to the low degree of complexity of the tissues.

6. When more than three animals are used, keep the Petri containing brains and DPBS in the refrigerator or on ice until SVZ dissection.
7. To prevent tissue and/or cells from sticking to the walls of the tip, rinse it several times with medium, before every dissociation step, or when a new tip is used.
8. It is preferable not to use the vacuum in this first step, to avoid the risk of sucking out the tissue.
9. Slightly tilt the pipetman, and press tip against the bottom of the tube to generate a fair amount of resistance.
10. Tissues from two or three mice are usually pooled to start a culture. Tissue from single mice can be similarly used if necessary, paying attention to the plating density.
11. A reliable cell count from adult tissue is sometimes difficult because of the presence of debris and the small number of cells that can be isolated. In the authors' experience, this protocol should yield about $0.5–1 \times 10^5$ cells from the subventricular tissue of one brain. Thus, if no careful quantification of the primary NSC number has to be carried out, cell suspension derived from one SVZ may be plated in one well of a 6-multiwell tissue culture plate in 5 mL final volume, yielding an approximate final cell density of about 2,500–5,000 cells/cm^2 (or 5–10 cells/μL).
12. Slightly tilt the pipetman, and press tip against the bottom of the tube to generate a fair amount of resistance.
13. Viability (number of live cells/total number of cells) after dissociation should never fall below 50–60 %.
14. If spheres are too small when passaged (Fig. 2b), the cell yield will be low; if spheres are allowed to grow too much before being passaged (Fig. 2c), the number of dead cells inside the spheres will be high, dissociation will be difficult, and viability of the culture will be very low.
15. Sometimes cells can adhere to the plastic of the culture dishes. It may depend on the mouse strain of origin: in our experience SVZ cells derived from FVB and C57/BL6 mice are more prone to adhesion than cells derived from CD1 mice. Usually adherent cells can be dislodged by gently tapping the flask.
16. By performing population analysis in long-term neurosphere cultures, we observe an overall enrichment and expansion of the stem population. This is likely due to the increased proportions of symmetric proliferative divisions over symmetric differentiative and asymmetric divisions [12], a mechanism

driven by the positive selection induced by the NSA culture conditions within the relatively small stem cell population [9, 13].

17. Thaw Matrigel stock solution at 4 °C. Use only pre-cooled sterile plastic pipette to aliquote Matrigel. Prepare 1 mL aliquotes in sterile cryovials and store them at −20 °C.
18. To prepare Matrigel solution, pre-cool 2 mL and 10 mL plastic pipettes at −20 °C and thaw one Matrigel vial at 4 °C or on ice. Filter 0.22 μm 100 mL of cold DMEM. Using a cold 2 mL plastic pipette take 1 mL of cold DMEM and then with the same pipette collect Matrigel from the vial and place it in a 50 mL sterile plastic tube. Using the same pipette, wash the vial with cold filtered DMEM and add this DMEM to the same 50 mL tube. Add 10 mL of cold filtered DMEM to the tube and mix well by pipetting the solution up and down. Add the obtained solution to the filtered DMEM and store it at 4 °C.
19. Do not filter Matrigel solution.
20. Spheres dimension is critical, as in **Note 14**.
21. To increase homogeneity in the cell population use spheres that have been subcultured at least for four passages, so that, in the starting cultures, short-term dividing precursors are absent. Generally, an increase in subculturing passages does not affect the proportion of the different cell types produced by stem cell progeny, upon differentiation. A shift towards glial fate in adult vs. neonates can be observed [14].
22. If dissociation has been efficient, almost the totality of the cells plated after subculturing should be single cells, which eventually proliferate in response to GF(s). If cultures are harvested after cells have made only one or two proliferation cycles (when one sees mostly doublets or very small clusters), culture will be highly enriched in stem-cell elements and very undifferentiated precursors.
23. Matrigel solution can be collected sterile and used a second time.
24. Seeding density is crucial for an efficient differentiation protocol: low cell density will result to poor yield of neurons and oligodendrocytes, whereas an higher cell concentration will result in massive cell death and eventually the detachment of the cell layer from the coverslip.
25. Under our culture conditions, simultaneous detection of the three cell phenotypes is usually successful at 7 days after plating. For immunofluorescence analysis using lineage-specific markers *see* refs. 14, 15.
26. Quantitative analysis of the relative proportion of neurons, astrocytes, and oligodendrocytes is the easiest method for comparing different protocols aimed to obtain NSC differentiation.

27. The protocol described for bulk cultures can be applied to differentiate single neurospheres.
28. Isopropanol can be used up to five freezing cycles.
29. In our experience, glycerol yields poor results when freezing NSCs.
30. For a good viability yield at thawing, it is crucial that neurospheres are intact at the moment of freezing. Never dissociate neurospheres before freezing. Moreover, do not let spheres grow too large before harvesting for cryopreservation, because it will reduce dramatically cell viability at thawing.

References

1. Kilpatrick TJ, Bartlett PF (1993) Cloning and growth of multipotential neural precursors: requirements for proliferation and differentiation. Neuron 10:255–265
2. Murphy M, Drago J, Bartlett PF (1990) Fibroblast growth factor stimulates the proliferation and differentiation of neural precursor cells in vitro. J Neurosci Res 25:463–475
3. Palmer TD, Ray J, Gage FH (1995) FGF-2-responsive neuronal progenitors reside in proliferative and quiescent regions of the adult rodent brain. Mol Cell Neurosci 6:474–486
4. Vescovi AL et al (1993) bFGF regulates the proliferative fate of unipotent (neuronal) and bipotent (neuronal/astroglial) EGF-generated CNS progenitor cells. Neuron 11:951–966
5. Weiss S et al (1996) Multipotent CNS stem cells are present in the adult mammalian spinal cord and ventricular neuroaxis. J Neurosci 16: 7599–7609
6. Reynolds BA, Tetzlaff W, Weiss S (1992) A multipotent EGF-responsive striatal embryonic progenitor cell produces neurons and astrocytes. J Neurosci 12:4565–4574
7. Reynolds BA, Weiss S (1992) Generation of neurons and astrocytes from isolated cells of the adult mammalian central nervous system. Science 255:1707–1710
8. Loeffler M, Potten C (1997) Stem cells—chapter: stem cells and cellular pedigrees—a conceptual introduction. Academic, London, UK
9. Galli R et al (2002) Emx2 regulates the proliferation of stem cells of the adult mammalian central nervous system. Development 129:1633–1644
10. Singec I et al (2006) Defining the actual sensitivity and specificity of the neurosphere assay in stem cell biology. Nat Methods 3:801–806
11. Reynolds BA, Rietze RL (2005) Neural stem cells and neurospheres–re-evaluating the relationship. Nat Methods 2:333–336
12. Morrison SJ, Kimble J (2006) Asymmetric and symmetric stem-cell divisions in development and cancer. Nature 441:1068–1074
13. Biffi A et al (2004) Correction of metachromatic leukodystrophy in the mouse model by transplantation of genetically modified hematopoietic stem cells. J Clin Invest 113: 1118–1129
14. Gritti A et al (2009) Effects of developmental age, brain region, and time in culture on long-term proliferation and multipotency of neural stem cell populations. J Comp Neurol 517:333–349
15. Cavazzin C et al (2006) Unique expression and localization of aquaporin- 4 and aquaporin-9 in murine and human neural stem cells and in their glial progeny. Glia 53(2):167–81

Chapter 4

Culturing and Expansion of Precursor Cells from the Adult Hippocampus

Dhanisha J. Jhaveri, Boris W. Prosper, and Perry F. Bartlett

Abstract

It is now well established that a resident population of neural precursor cells continues to generate new neurons in the adult hippocampus throughout life. Numerous studies have suggested that these newborn neurons preferentially participate in the functional hippocampal circuitry that leads to enhancement of learning, cognition and mood. Therefore, understanding the molecular mechanisms that regulate the activity of these endogenous precursor cells is paramount to develop novel regenerative strategies for the treatment of neurological and psychiatric disorders. The neurosphere assay has been instrumental in discovering the presence of stem and precursor cell population from several brain regions. In this chapter, we describe this assay to specifically isolate and culture neural stem and precursor cell populations from the adult hippocampus of mice. In addition, we provide methods to conduct detailed assays to examine their functional properties such as proliferation, self-renewal, and differentiation.

Key words Hippocampus, Neural precursor cells, Neural stem cells, Neurogenesis, Neurospheres, Adult, Mouse

1 Introduction

In the adult hippocampus, neurogenesis occurs in a highly orchestrated fashion, beginning with the activation and proliferation of the neural precursor cells, which subsequently differentiate to generate new neurons [1]. After birth, approximately half of these new neurons survive beyond 4–6 weeks and functionally integrate into the hippocampal circuitry. Interestingly, this precursor pool is "plastic" and responds to a number of molecular and environmental factors such as neuronal activity, exercise, stress, and antidepressants, which in turn regulate the production of new neurons [2, 3]. Understanding the mechanisms that regulate the generation of functional neurons will be important for the design of future regenerative therapies to combat both neurological and psychiatric disorders. As interest from the neuroscience community grows, efforts across a range of institutions are currently underway to fully

Brent A. Reynolds and Loic P. Deleyrolle (eds.), *Neural Progenitor Cells: Methods and Protocols*, Methods in Molecular Biology, vol. 1059, DOI 10.1007/978-1-62703-574-3_4, © Springer Science+Business Media New York 2013

elucidate the functional capacity of hippocampal precursor cells. The lack of a definitive marker to identify and follow this precursor pool has made it difficult to examine the functional properties of these cells in vivo. Often, hippocampal precursors are identified based on the co-expression of a glial cell marker, GFAP, and neurofilament nestin; however, expression of these markers is not unique to precursor cells.

To date, an in vitro neurosphere assay has emerged as one of the most powerful and widely used tools to identify, expand, and quantify the precursor cell populations from various regions of the developing and adult brain [4]. This assay is based on one of the fundamental properties of precursor cells: the ability to proliferate extensively in the presence of growth factors such as EGF and bFGF. Using this assay it is possible to generate a three-dimensional ball of cells called a neurosphere. In this system, a *bona fide* stem cell displays large proliferative potential, extensive self-renewal capacity, and differentiation potential. These neural stem cells can be distinguished from the precursor cell population, which typically has limited proliferative capacity. The reproducibility and the relative ease of performing and quantifying have made the neurosphere assay an attractive tool in the discovery of the genetic and environmental factors that regulate the activity of the neural precursor cells.

Until recently, whether the adult hippocampus harbored a true stem cell pool, similar to that found in the subventricular zone (SVZ), was under contentious debate. A previous study found that while it was possible to obtain neurospheres from the hippocampus of an adult mouse, these neurospheres did not have the capacity to self-renew, suggesting that only a precursor cell population with limited proliferative potential existed in the adult hippocampus [5]; however, a recent landmark study showed that a large proportion of the hippocampal precursors were in fact quiescent and could only be activated by neural activity [6]. More importantly, in the presence of depolarizing levels of potassium that mimics neural activity in vitro, a small pool of true stem cells was activated that displayed all of the hallmark properties of stem cells. Since then, another latent population has been identified that could be activated by norepinephrine, a member of the monoamine family of neurotransmitters [7]. By performing the neurosphere assay at a clonal density, i.e., one cell per well, this study demonstrated that the effect of norepinephrine on the stem and precursor cell population was direct and mediated via the β3 adrenergic receptor subtype. In addition, this assay has been instrumental in uncovering that a large pool of latent precursors exists in the aged hippocampus, which can be activated by appropriate neural activity.

In this chapter, we describe, in detail, methods to isolate and culture both proliferating and latent populations of stem and precursors from the adult hippocampus. In addition, we provide detailed assays that examine the functional properties of these precursor populations.

2 Materials

2.1 Tissue Culture Medium and Stocks

1. NS Basal Medium: 1 L DMEM-F12, 3.75 g glucose, 1.125 g sodium bicarbonate, 5 mL of 1 M HEPES solution. Combine in a large vessel and add MilliQ water to a final volume of 9 L. Adjust pH to 7.2, filter sterilize and store at 4 °C.
2. NeuroCult Supplement (StemCell Technologies).
3. Bovine serum albumin (BSA).
4. HEM (Minimum Essential Medium with HEPES): 1 L MEM, 17.5 mL Penicillin/streptomycin (stock: Penicillin-10,000 units/mL, Streptomycin-10,000 μg/mL), 16 mL of 1 M HEPES solution. Mix all ingredients together in MilliQ water to a final volume of 8.75 L. Adjust pH to 7.2, filter sterilize, and store at 4 °C.
5. Heparin stock: Dissolve 555.56 mg Heparin (Sigma) in 147 mL of tissue grade quality water. Filter sterilize and store at −30 °C.
6. 0.1 % papain mix: 5 mL of 2 mg/mL Papain (Worthington), 2.5 mL of 0.1 % of DNase I (Roche) in HEM, 2.5 mL Hank's Balanced Salt Solution (HBSS, Thermo).
7. EGF (Epidermal Growth Factor) stocks: Dissolve 100 μg EGF (BD Biosciences) in 9 mL of NS Media (*see* NS Media recipe), 1 mL of Proliferation Supplement. Aliquot and store at −30 °C.
8. bFGF (Basic Fibroblast Growth Factor, Bovine) stock: Dissolve 10 μg bFGF (Roche) in 0.9 mL NS Media, 0.1 mL Proliferation Supplement, and 20 μL of 10 % BSA solution. Aliquot and store at −30 °C.
9. Penicillin/Streptomycin (Gibco).

2.2 Dissection Tools

1. Two curved forceps, with tip dimensions of 0.5 × 0.04 mm (Fine Science Tools).
2. Sterile disposable surgical scalpels.
3. Sterile Petri dishes (BD falcon, 100 × 15 mm).

2.3 Primary Neurosphere Culture and Passaging

1. 0.1 % Papain mix (*see* **Note 1**).
2. NSA Medium preparation (*see* Table 1 and **Note 2**).

2.4 Neurosphere Differentiation and Immunocytochemistry

1. Poly-D-lysine-coated BioCoat eight-well culture slides (BD Biosciences), or, coat plastic chamber slides with poly-ornithine (15 % in PBS).
2. Differentiation medium (*see* Table 2).

Table 1
NSA medium preparation

	50 mL	100 mL	150 mL	200 mL
NS Basal Medium	44 mL	88 mL	132 mL	176 mL
NeuroCult Supplement	5 mL	10 mL	15 mL	20 mL
BSA 10 %	1.34 mL	2.67 mL	4 mL	5.32 mL
Penicillin/Streptomycin	505 μL	1.1 mL	1.6 mL	2.2 mL
Heparin	50 μL	100 μL	200 μL	200 μL
EGF	100 μL	200 μL	300 μL	400 μL
bFGF	50 μL	100 μL	200 μL	200 μL

Table 2
Differentiation medium

	50 mL	100 mL
NS Basal Medium	45 mL	90 mL
NeuroCult Supplement	5 mL	10 mL
Penicillin/streptomycin	505 μL	1.1 mL

3. 4 % Paraformaldehyde (PFA) solution in 0.1 M PBS.
4. PBS-T solution: 0.1 % Triton 100X in 0.1 M PBS.
5. Primary antibodies- Neurons: βIII tubulin (1:2,000; Promega), astrocytes: glial fibrillary acidic protein (GFAP; 1:500; DakoCytomation) and oligodendrocytes: myelin basic protein (MBP; 1:500; Millipore).
6. Nuclear stain: 4′,6′-diamidino-2-phenylindole (DAPI; 1:5,000).
7. DAKO fluorescent mounting medium.
8. Glass coverslips (22 × 64 mm).

3 Methods

3.1 Isolation of the Hippocampus from the Adult Mouse Brain

1. Sacrifice the mice by cervical dislocation and quickly but carefully remove the brains (*see* **Note 3**). Keep each brain in a 50 mL tube containing ice-cold HEM medium.
2. Under aseptic conditions (in laminar flow), using a #10 surgical blade, make a clean coronal cut along optic chiasm to remove the rostral region of the septo-diencephalon (olfactory bulbs and SVZ; *see* Fig. 1).

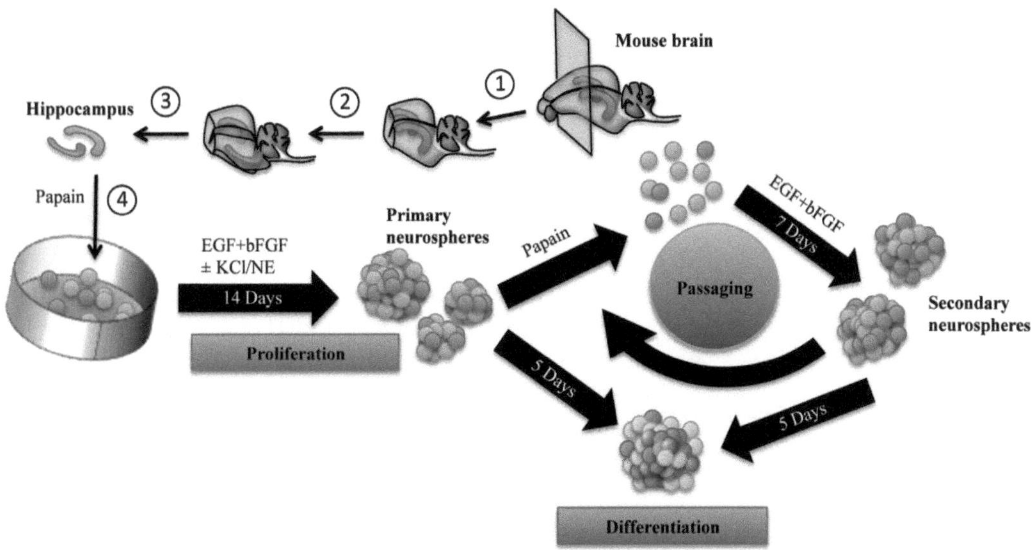

Fig. 1 The neurosphere assay is a method that allows for the determination of the proliferation potential, capacity for self-renewal (passaging), and multipotency (differentiation) of activated precursors. By measuring these parameters we are able to analyse the precursor/stem cells activity in the hippocampus. As each neurosphere is clonally derived from a single cell, the size, the number of passages, and the differentiation of the neurosphere provide information about the precursor cell. 1–4: steps of hippocampal removal from rodent brain (describe in Subheading 3.1)

3. Using the forceps, push each of the hemispheres aside to gain access to the hippocampus. Untie both caudal and rostral extremities of hippocampus and, by gentle circular movements, raise it from the overlying cortex. Then, with the edge of the forceps, cut out surrounding cortical tissue and carefully remove the hippocampus. While dissecting the remaining hippocampi, keep the already isolated hippocampus in a drop of ice-cold HEM in a petri dish, keeping the dish on ice until all dissections are complete (*see* **Note 4**).

3.2 Neurosphere Assay

Perform the following procedures in a tissue culture (TC2) certified hood.

1. Using a scalpel blade, gently mince the hippocampi in the petri dish until no obvious lumps of tissue can be seen.
2. Add 1 mL of papain mix to the minced hippocampus and transfer the mixture to a 15 mL tube (two hippocampi per tube) using 1 mL pipette.
3. Place tubes in a 37 °C water bath for 16 min to allow papain digestion. At 8 min, remove tubes from the water bath and gently triturate (*see* **Note 5**) the mixture 3–4 times using a 1 mL pipette and return the tubes to the 37 °C water bath for the remaining time.

4. At the end of the 16 min digestion, gently triturate 2–3 times and add 2 mL of NSA.
5. Centrifuge 15 mL tube at 100 × *g* for 5 min at 22–25 °C.
6. Carefully remove the supernatant and resuspend the pellet in 2 mL of NSA.
7. Centrifuge, as per **step 5**, before resuspending in 1 mL of NSA.
8. Use 10 μL of suspension for cell and viability count (add 10 μL of Trypan blue to 10 μL of suspended cells and count the live (bright) and dead (blue) cells using a haemocytometer; (*see* **Note 6**). The cell suspension should be kept on ice during the counting to minimize cell death.
9. Following the cell count, divide the hippocampal cell suspension equally into 50 mL tubes and add 9.5 mL of NSA to each of the tubes. Keep one hippocampal cell suspension as the control and treat the other with either KCl (15 mM) or norepinephrine (10 μM).
10. Plate each of the control/treatment conditions into 48 wells (i.e., each treatment occupies half of 96-well plate) with 200 μL per well.
11. Place the plates in a humidified incubator at 37 °C with 5 % CO_2 and culture for 14 days.
12. On day 14, count the total number of primary neurospheres using an inverted brightfield microscope (10× objective, *see* **Note 7**). A systematic categorization of neurospheres (i.e., neural stem cell-derived or precursor cell-derived) can be done by measuring the diameter of each neurosphere using an eyepiece graticule. The neurospheres that measure >200 μm in diameter are generally derived from hippocampal stem cells, whereas those that measure between 40 and 200 μm are derived from proliferation of neural precursor cells (Fig. 2).

3.3 Neurosphere Differentiation

1. Following quantification, collect the neurospheres from all wells for each treatment group using 1 mL pipette and place them into a 15 mL tube (one tube per treatment group).
2. Let the tube stand for 5 min at room temperature to allow neurospheres to settle (*see* **Note 8**).
3. Gently remove the supernatant either manually (using 1 mL pipette) or using vacuum-aided suction, ensuring that ~300 μL of the medium containing the neurospheres remains at the bottom of the tube.
4. Resuspend the neurospheres in 2 mL of fresh differentiation medium.

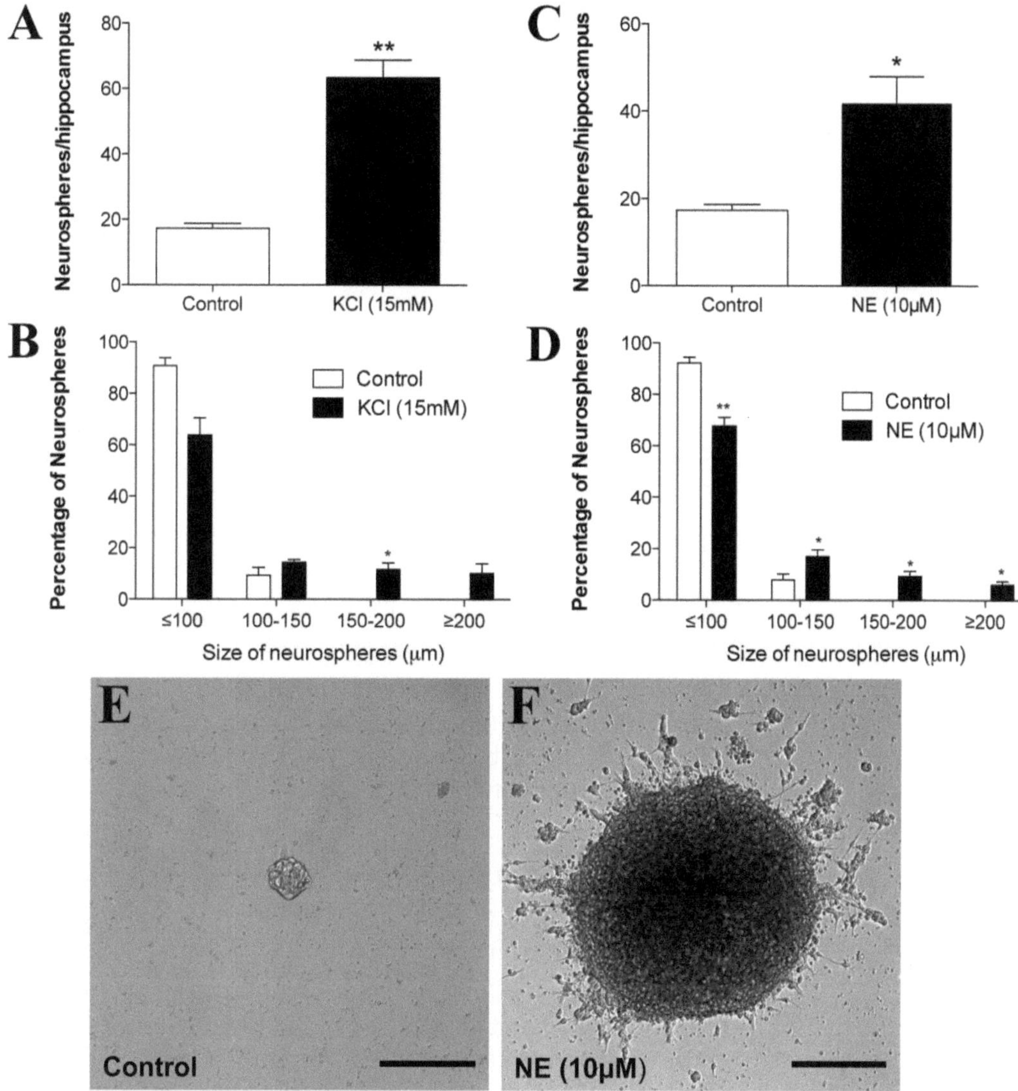

Fig. 2 KCl and norepinephrine (NE) activate latent precursor cell populations from the adult hippocampus. (**a**, **b**) Treatment of adult hippocampal cells with both KCl (15 nM) and NE (10 μM) in the presence of bFGF and EGF significantly increased neurosphere formation by up to two- and threefold, respectively ($n=3$; mean ± SEM; $^*p<0.05$, $^{**}p<0.01$). In addition, the neurospheres obtained in both treatments were significantly larger (**b**–**d**), with an increase in the percentage of neurospheres measuring 100–150, 150–200, and >200 μm observed ($n=3$; mean ± SEM; $^*p<0.05$, $^{**}p<0.01$). (**e**, **f**) Control and norepinephrine-stimulated neurospheres. Unstimulated precursor cells show modest proliferative activity and give rise to small neurospheres (**e**, <100 μm). In contrast, when stimulated with 10 μM of NE, an activated population of precursor cells displays high proliferative activity, thereby giving rise to a large neurosphere (**f**, measuring ~450 μm). Scale bar = 200 μm

5. Repeat **steps 2** and **3** at least two times to wash away any residual NSA medium.
6. Resuspend the neurospheres in 500 μL of differentiation medium (*see* **Note 9**).
7. Transfer the 500 μL neurosphere suspension to one of the wells of a poly-D-lysine-coated BioCoat eight-well culture slide.
8. Place the slide in a sterile petri plate and incubate it in a humidified incubator for 5 days at 37 °C with 5 % CO_2.

3.4 Neurosphere Immunocytochemistry

Perform all washes on an orbital shaker (40–50 rpm) for 10–15 min at room temperature.

1. On day 6, gently remove the differentiation medium from each well using a 1 mL pipette. Fix the neurospheres by adding 300 μL of ice-cold 2 % PFA for 30 min at 4 °C.
2. Following fixation, wash the neurospheres three times with PBS-T.
3. Block the neurospheres in PBS-T with 3 % Normal Goat Serum for at least 60 min at room temperature.
4. Remove the blocking solution and add 250 μL of a primary antibody solution (antibody stock diluted to the appropriate concentration in blocking solution) to each of the wells. Incubate overnight at 4 °C.
5. The next day, wash the neurospheres three times with PBS-T.
6. Incubate the neurospheres in 250 μL of secondary antibody solution for 2–4 h at room temperature.
7. Wash the neurospheres twice with PBS-T before a final wash with PBS.
8. Remove the chamber grid as per the instruction of the supplier. Cover the slide with DAKO mounting medium and gently lower the glass coverslip onto the slide, ensuring no air bubbles are trapped.
9. Seal the edge of the coverslip with nail polish and store the slides at 4 °C.
10. Observe them under an appropriate miscroscope fitted with optics suitable for capturing fluorescence (Fig. 3).

3.5 Passaging a Single Adult Hippocampal Neurosphere

1. Mark the wells that have neurospheres measuring >200 μm in diameter, making sure there is only a single sphere per well of the 96-well plate. If there is more than one neursophere per well, dilute the content of the wells over five wells such that there is only one neurosphere per well.

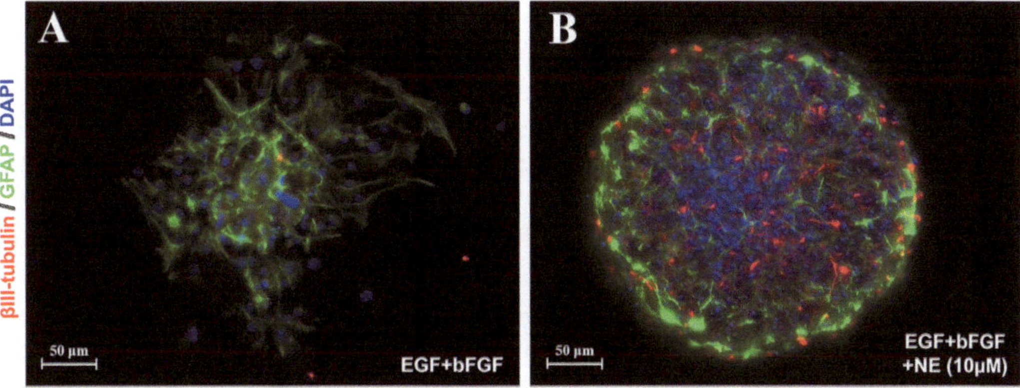

Fig. 3 Norepinephrine-treated precursors are highly neurogenic. Neurospheres from both the control and experimental condition (NE) were labeled with the glial marker GFAP (*green*) and the neuronal marker β3-tubulin (*red*). (**a**) Under control conditions, neurospheres mainly differentiated into glial cells; however, in the presence of NE (10 µM), majority of the differentiated neurospheres contained a large number of β3-tubulin-positive neurons (**b**)

2. Choose appropriate control neurospheres to passage simultaneously.
3. Carefully remove 180 µL of medium from the well.
4. Add 100 µL of Trypsin–EDTA mix and incubate at room temperature for 3 min.
5. Add 100 µL of trypsin inhibitor to the well and triturate the mix at least 6–8 times using a 200 µL pipette.
6. Transfer the triturated mix to a single well of a 24-well plate containing 2 mL of complete NSA medium. Incubate the plate for 7 days at 37 °C with 5 % CO_2.
7. On day 8, repassage the wells containing secondary neurospheres by collecting all the neurospheres from individual wells and placing them in 15 mL tubes. Centrifuge at $100 \times g$ for 5 min. Carefully remove the supernatant and treat the pellet containing neurospheres with 500 µL of Trypsin–EDTA mix. Incubate at 37 °C for 5 min and then add an equal volume of trypsin inhibitor. Centrifuge and remove the supernatant. Add 1 mL of NSA and gently triturate 6–8 times. Plate all the cells in 2 mL of NSA, again in a single well of a 24-well plate. Incubate the plate for 7 days at 37 °C with 5 % CO_2.
8. During the third passage, repeat the **step** 7. After trituration, perform cell and viability count (as per **step 8** of neurosphere assay) using a hemocytometer. Plate 2×10^5 cells in a T25 flask with 5 mL NSA medium. Complete a further ten rounds of passaging to assess self-renewal property of the initial (primary) stem or precursor cells.

4 Notes

1. It is best to first defrost and then pre-heat the papain mix to 37 °C so as to activate the enzyme, which can then efficiently break down the extracellular matrix in the dissociated hippocampal tissue.
2. NSA medium should be freshly prepared and is good to use for up to 3 days if stored at 4 °C.
3. A steady but significant decrease in hippocampal precursor and stem cells activity has been observed after 10 weeks of age in C57BL6/J mice [6]. Thus, is preferable to use 8–10-week-old mice to culture hippocampal neural stem and precursor cells. It is also advisable to use either all female or all male mice for experiments that are under a single project to minimize sex-specific variability in the neurosphere numbers.
4. Excess cell death can be avoided by minimizing the dissection time. This can be achieved by using maximum of four mice at a time for hippocampal dissection.
5. While triturating, it is important to dissociate all of the tissue clusters. Trituration is done by gently pipetting up and down using a 1 mL pipette, with the tip touching the bottom of the tube. Make sure not to triturate more than 3–4 times per tube, as excessive trituration may lead to precursor cell death. On the other hand, if trituration is not done properly, a good single cell suspension cannot be obtained and small clusters of hippocampal tissue may hinder proliferation of hippocampal precursors.
6. An average cell count for two hippocampi from a single brain is $3.54 \pm 0.08 \times 10^5$ cells (mean ± SEM, $n = 10$).
7. General variability in terms of the neurosphere number obtained per brain (from two hippocampi) is often observed, which primarily depends on the strain of the mice and the dissection technique.
8. Avoid centrifugation as this may lead to clumping of spheres. Again, before taking out the supernatant, check the bottom of the tube to verify the presence of neurospheres.
9. If there are numerous neurospheres, it is preferable to dilute them in 1 mL differentiation medium and split them into two to three chambers in differentiation slide, so as to avoid their overlap during the process of differentiation.

References

1. Ming GL, Song H (2005) Adult neurogenesis in the mammalian central nervous system. Annu Rev Neurosci 28:223–250
2. Vukovic J et al (2011) Activation of neural precursors in the adult neurogenic niches. Neurochem Int 59:341–346
3. Ma DK et al (2009) Adult neural stem cells in the mammalian central nervous system. Cell Res 19:672–682
4. Reynolds BA, Weiss S (1992) Generation of neurons and astrocytes from isolated cells of the adult mammalian central nervous system. Science 255:1707–1710
5. Bull ND, Bartlett PF (2005) The adult mouse hippocampal progenitor is neurogenic but not a stem cell. J Neurosci 25:10815–10821
6. Walker TL et al (2008) Latent stem and progenitor cells in the hippocampus are activated by neural excitation. J Neurosci 28:5240–5247
7. Jhaveri DJ et al (2010) Norepinephrine directly activates adult hippocampal precursors via β3-adrenergic receptors. J Neurosci 30:2795–2806

Chapter 5

Isolate and Culture Neural Stem Cells from the Mouse Adult Spinal Cord

Jean-Philippe Hugnot

Abstract

Whereas neural stem cells and their niches have been extensively studied in the brain, little is known on these cells, their environment and their function in the adult spinal cord. Adult spinal cord neural stem cells are located in a complex niche surrounding the central canal and these cells expressed genes which are specifically expressed in the caudal central nervous system (CNS). In depth characterization of these cells in vivo and in vitro will provide interesting clues on the possibility to utilize this endogenous cell pool to treat spinal cord damages. We describe here a procedure to derive and culture neural spinal cord stem cells from adult mice using the neurosphere method.

Key words Spinal cord, Neural stem cells, Neurospheres, Niche

1 Introduction

Much attention has been given to stem and progenitor cells in the brain, whereas little is known about these cells in the spinal cord. The persistence of stem cells in this caudal region of the CNS was reported using adherent and non-adherent culture conditions in the late 1990s [1, 2]. The neurosphere assay was, however, instrumental in their discovery, as this assay is particularly suited to demonstrate, at the clonal level, the cardinal properties of stem cells, i.e., multipotentiality, self-renewal, and extended proliferation capabilities. Indeed, in 1996 Weiss et al. reported that in mice, isolated thoracic and lumbar spinal cord cells, grown in the presence of FGF2 and EGF, were able to form multipotent and passageable neurospheres [2]. Using microdissection and cytometric analysis, these cells were located primarily in the central canal region [3–5]. Progenitor cells with a more limited proliferation potential are also present in the parenchyma [3, 5–8]. Spinal cord neurospheres have different properties from those derived from the brain [9]. In addition, neurospheres derived from different parts of the spinal cord have different growth and differentiation properties [10]. At the molecular

Brent A. Reynolds and Loic P. Deleyrolle (eds.), *Neural Progenitor Cells: Methods and Protocols*, Methods in Molecular Biology, vol. 1059, DOI 10.1007/978-1-62703-574-3_5, © Springer Science+Business Media New York 2013

level, neurospheres that are derived from the cervical, thoracic, and lumbar regions maintain the expression of specific and different rostro-caudal combinations of developmental homeogenes of the Hox family [5]. These data indicate that even after several passages in vitro, these cells maintain molecular cues from their original position.

The spinal cord stem cells are situated in a complex niche which contains different cell types and which maintains the expression of developmental signalling pathways and genes in an adult environment [11]. This niche shows both similarities and dissimilarities with the brain niches. In adults, the niche is mostly in a dormant state but can be reactivated by physical exercise, by traumatic lesions, and by degenerative diseases such as multiple sclerosis and amyotrophic lateral sclerosis [11]. Importantly, in vitro, these cells appear to remain competent to respond to various morphogens which can direct their differentiation toward electrophysiological active motoneurons [12].

Collectively, the adult mouse spinal cord niche and stem cells represent an attractive model for getting insights on the molecular cues which allow the maintenance of stem cells in adults. A better description of the properties of this endogenous cell pool is certainly a step toward the design of rational strategies to treat spinal cord damages. Here we describe a procedure to derive and culture neurospheres from the adult mouse spinal cord.

2 Materials

2.1 Spinal Cord Dissection

1. EtOH 70 % for sterilization of instruments.
2. Spray bottle with EtOH 70 %.
3. PBS (phosphate buffer saline) or HBSS (Hanks Balanced Salt Solution) for rinsing of instruments.
4. Sterile dissection hood.
5. Petri dishes for dissection.
6. Gloves sprayed with EtOH 70 %.
7. Dissection scissors for bones (for instance "toughcut" scissors from Fine Science Tools company).
8. Sharp spring scissors—for instance 10 mm Blades Moria MC26 Dowell (straight or angled up) or No 15024-10 from Fine Science Tools company.
9. Standard forceps (for instance Dumond no.7).
10. Fine scissors for dissection.

2.2 Enzymatic Dissociation of Spinal Cord

1. Hyaluronidase (Sigma H3884 or H4272) (7 mg/ml). Weigh 100 mg and add 14 ml PBS, sterilize by filtration, aliquot in 500 μl, store at −20 °C.

2. Trypsine (Sigma T4799) (13 mg/ml). Weigh 133 mg and add 10 ml PBS, sterilize by filtration, aliquot in 500 µl, store at -20 °C.
3. DNAse I 10 mg/ml (Roche 11284932001, 2,000 units/mg). Weigh 100 mg and add 10 ml sterile water, sterilize by filtration, aliquot in 500 µl, store at -20 °C.
4. HBSS (Hanks Balanced Salt Solution) w/o and with Ca^{2+} and Mg^{2+}.
5. Sucrose (0.9 M) (Sigma S7903): Weigh 15.4 g and dissolve in HBSS w/o Ca^{2+} and Mg^{2+}. Complete to 50 ml, sterilize by filtration. Store at 4 °C or -20 °C.
6. RBC solution (optional): solution to eliminate red blood cells. This solution is important if the cellular suspension is to be used for cytometry. Weigh 415 mg NH_4Cl, 2 mg Na2EDTA, 50 mg $KHCO_3$, H_2O for 50 ml final. This solution is generally prepared fresh for each experiment.
7. 40 µm cell strainer (BD 352340) for 50 ml disposable tubes.
8. Sterile microfuges and 15 ml tubes.

2.3 Spinal Cord Neurosphere Culture

1. Media: 500 ml DMEM/F12 (In Vitrogen 21331020)) without glutamine with 0.3 % glucose, $NaHCO_3$, and pyruvate, 5 ml N2 serum replacement, 5 ml sterile L-Glutamine 200 mM, 7.5 ml sterile glucose (cell culture grade) 200 g/l, 1 ml of insuline (see below), 100 µl of ciprofloxacine (antibiotics, Euromedex) at 10 mg/ml (dissolved in water), 100 µl of gentamycine at 50 mg/ml (In Vitrogen). For high density culture, the addition of B27 serum replacement (10 ml for a 500 ml bottle) is not necessary however it may improve the growth rate. This completed media can be kept at 4 °C for at least 2 weeks. We also add occasionally fungin (Invivogen, 50 µg/ml final) and fungizone (In Vitrogen, 0.25 µg/ml final) to prevent or eradicate fungi contaminations.
2. For clonal analysis, i.e., one cell per well of a 96-well plate, it is necessary to add B27 (without vitamin A, In Vitrogen 12587-010) in addition to N2. The reason why B27 addition is required is unclear. This contains antioxidant molecules such as catalase which may help cells resist the low density.
3. Insulin (Sigma, 100 mg, I1882 sterile). Add under the hood 10 ml of sterile H_2O acidified with HCl 10 mM directly into the vial. Leave for 30′ without shaking nor pipetting to allow slow dissolution. Aliquot in 1 ml at 10 mg/ml, store at -20 °C.
4. Growth factors and other components to be added in the flask (kept at –80 °C then at 4 °C for 15 days after thawing):

   Human FGF2 (Peprotech), stock solution: 50 µg/ml prepared in sterile buffer: HEPES 1 mM pH 7.4, 0.1 % BSA (Cristalline Gibco). Working concentration: 10 ng/ml.

Human EGF (Peprotech), stock solution: 50 μg/ml prepared in sterile buffer HEPES 1 mM pH 7.4, 0.1 % BSA (Cristalline Gibco). Working concentration: 10 ng/ml.

Heparin (Sigma H3149), stock solution: 2 mg/ml in H_2O, sterilized by filtration. Working concentration: 2 μg/ml.

5. Coating vessels with poly HEMA (anti-adhesive coating)

   Dissolve the poly HEMA (poly-2-hydroxyethyl methacrylate Sigma P3932) overnight in EtOH 95 °C under agitation to obtain a 10 mg/ml solution (*see* **Note 1**). We usually prepare a 200 ml bottle. Sterilize by filtration. Keep it at 4 °C for several months. To coat the dishes or flasks, apply the following volumes: 4 ml for a 75 cm^2 flask, 2 ml for a 25 cm^2 flask, 0.4 ml, 0.3 ml, 10 μl for 6-, 24-, 96-well plate, respectively.

   Put vessels in a dry incubator O/N (*see* **Note 2**) at 50–60 °C to dry the poly HEMA completely (*see* **Note 3**). The coated vessels can be kept at 4 °C for several months. Before using, rinse twice with sterile water or HBSS. It is a critical step as traces of free poly HEMA appear to be toxic for the cells.

6. Neurosphere dissociation for passage.
   (a) Trypsin 0.25 %/EDTA (In Vitrogen, 25200056).
   (b) Trypsine Inhibitor (Sigma T9003), prepare a solution at 5 % (50 mg/ml) in HBSS w/o Ca^{2+} and Mg^{2+}, then filter on 0.22 μm filter and store at −20 °C.
   (c) $CaCl_2$ 20 mM in water or PBS, sterilized by filtration.

3 Methods

3.1 Spinal Cord Dissection

1. Dissecting instruments are placed in a beaker with EtOH 70 % for sterilization and are rinsed with PBS or HBSS before use.
2. Kill the mice by CO_2 inhalation or other methods. Avoid cervical dissociation as it may damage the cervical spinal cord (*see* **Note 4**).
3. Place the animals in the dissection hood and spray the back with EtOH 70 %.
4. Cut the back skin with a scalpel and pull both sides apart (Fig. 1, 1).
5. With a tough scissors cut the caudal end of the vertebral column (Fig. 1, 2). then cut along both sides of the vertebral column up to the skull (Fig. 1, 3–6).
6. Remove the entire vertebral column from the body by cutting at the base of the skull (Fig. 1, 7, 8).

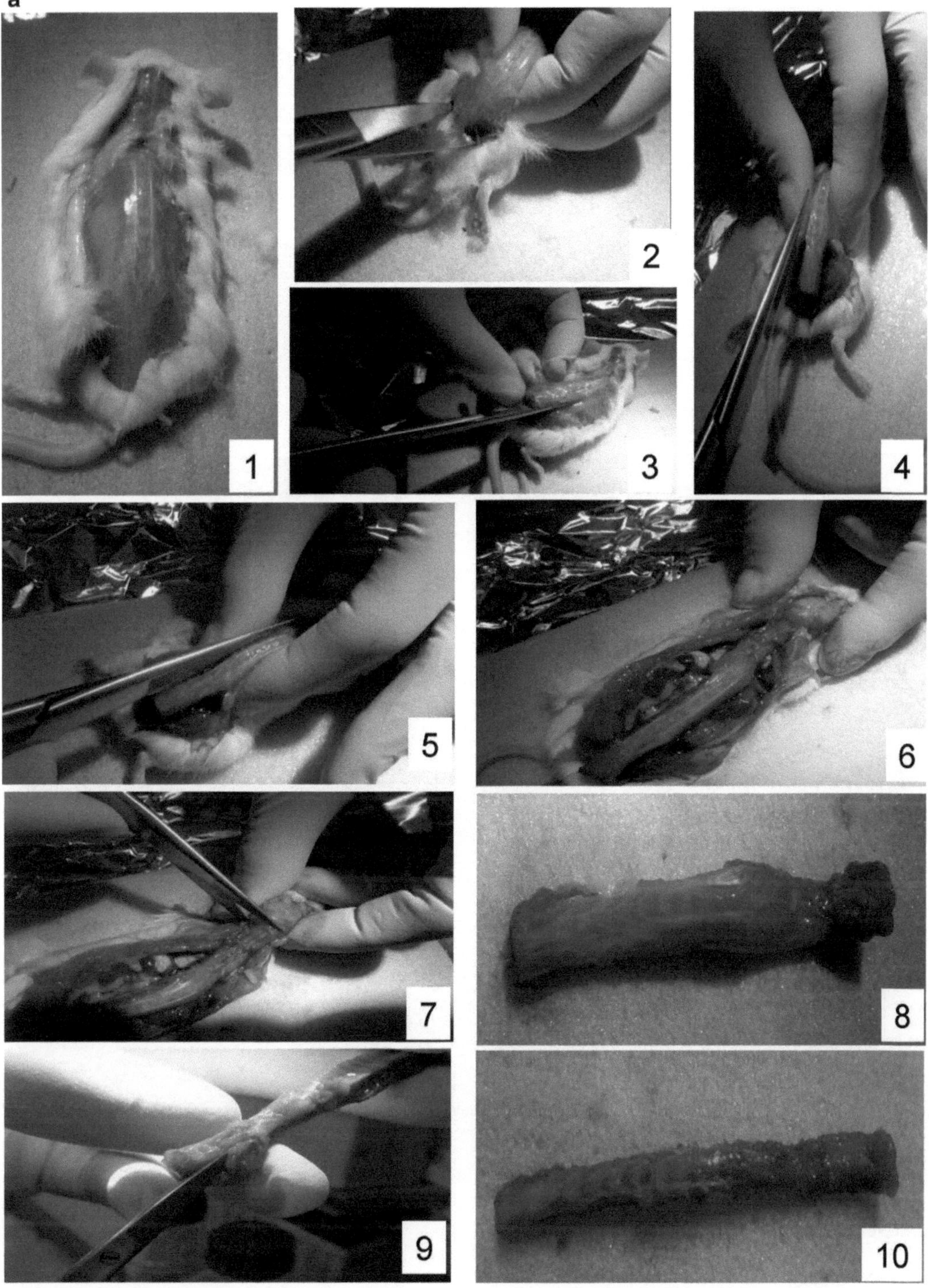

Fig. 1 Spinal cord dissection steps

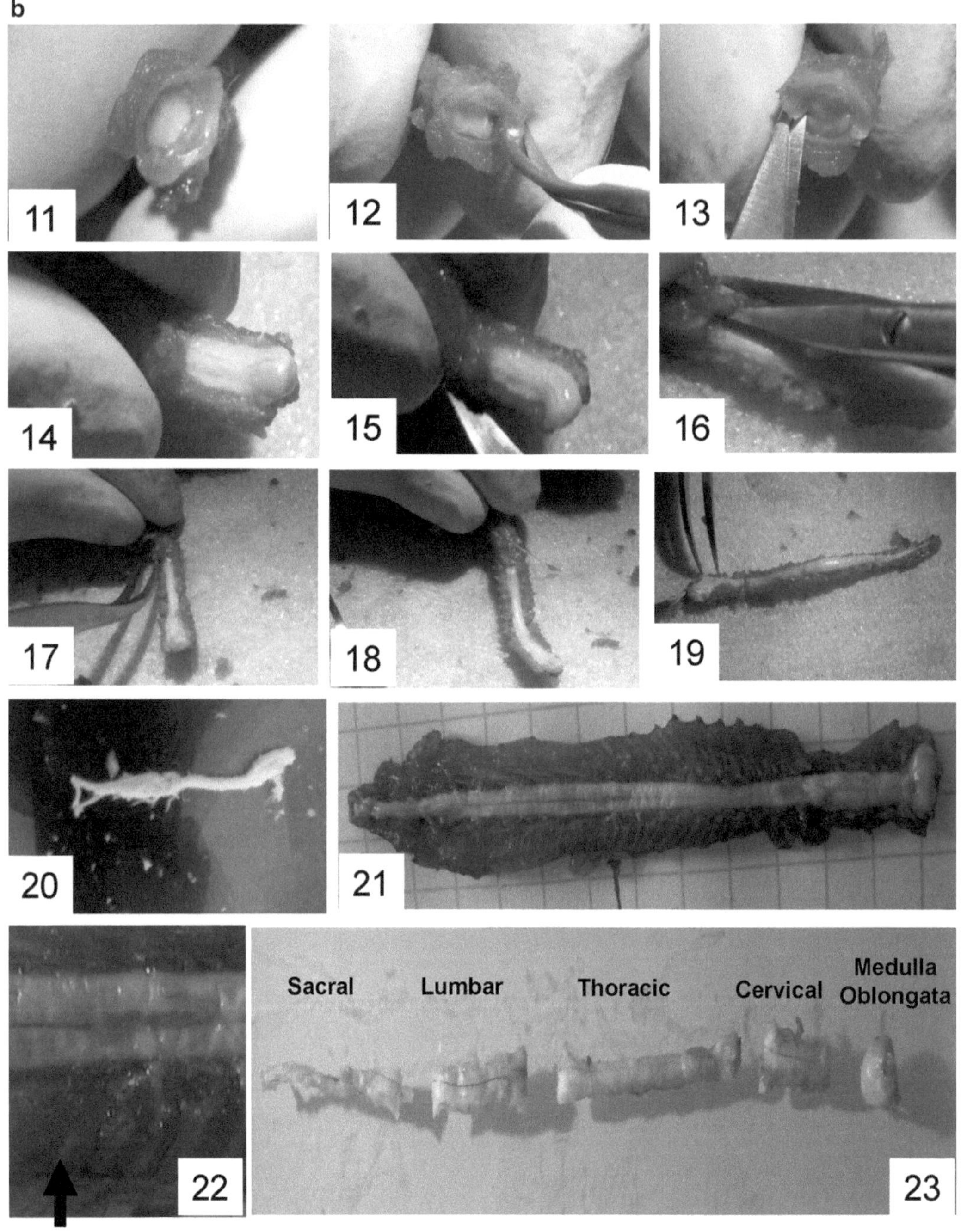

Fig. 1 (continued)

7. Holding the vertebral column in one hand (*see* **Note 5**), remove muscles and fat which are attached to the vertebral column as much as possible with the scissors (Fig. 1, 9, 10).
8. Beginning from the cervical region (Fig. 1, 11), insert the sharp spring scissors between the lateral small space between the spinal cord and the column and make a small cut (Fig. 1, 12). The same cut is made on the other side (Fig. 1, 13). Perform a laminectomy by removing the small piece of overlying bone so that a part of the spinal cord becomes visible (Fig. 1, 14). Repeat this step lengthwise down to the caudal region (Fig. 1, 15–19).
9. Gently removed the spinal cord from the vertebral column with forceps (Fig. 1, 20) and put it in a Petri dish with PBS or proceed to the dissociation step.
10. If different parts of the spinal cord are needed, the lower limit of the thoracic region is delimited by the last pairs of ribs (floating ribs) (Fig. 1, 21–23). The weights for these regions are approximately: cervical, thoracic, lumbar: 30, 44, 32 mg, respectively.

3.2 Cellular Preparation from Spinal Cord

1. Put the dissected spinal cord in a microtube containing 500 μl of HBSS with Ca^{2+} and Mg^{2+}.
2. Using fine scissors directly in the microtube, mince the spinal cord into very small pieces.
3. Add 130 μl hyaluronidase, 130 μl trypsin, 25 μl DNAse I.
4. Incubate for 30′ at 37 °C in a water bath. After 10′, 20′, and 30′ pipette up and down with a P1000 micropipette to homogenize.
5. Transfer to a 15 ml tube, add 6 ml of HBSS w/o Ca^{2+} and Mg^{2+}, centrifuge at 380 × *g* for 5 min (*see* **Note 6**).
6. Resuspend in 0.6 ml of HBSS w/o Ca^{2+} and Mg^{2+} and add 20 μl of trypsin inhibitor.
7. Dissociate the spinal cord fragment by gently pipetting up and down with a P1000 micropipette (*see* **Note 7**).
8. Fit a 40 μm cell strainer onto a 50 ml centrifuge tube.
9. Transfer the spinal cord suspension to the filter, let it go through and rinse with 2 or 3 ml HBSS w/o Ca^{2+} and Mg^{2+}.
10. Optional: To eliminate red blood cells, add RBC solution at a volume: volume ratio of 4:1 to 9:1. Place on ice for 10 min to allow red blood cells to lyse then proceed to **step 11**.
11. Centrifuge at 380 × *g* for 5 min and resuspend the pellet in 2 ml of sucrose solution.
12. Centrifuge at 750 × *g* (*see* **Note 6**) for 30′. Very carefully, remove the tubes from the rotor and place them on a rack, avoiding any abrupt movement.
13. Eliminate the white layer (myelin) by aspiration using a Pasteur pipette or a yellow tip connected to a vacuum pump (*see* **Note 8**).

14. Resuspend the pellet in 500 μl of complete medium and perform a cell count. The cell suspension is contaminated by debris and a lot of red blood cell. These are smaller and less refringent than the spinal cord cells (Fig. 2). These cells do not disturb the neurosphere growth and will be eliminated through the passages.
15. Around 150–300,000 cells are obtained by spinal cord. The cells are seeded in a 25 cm^2 flask with 5 ml of medium containing FGF2, EGF, and Heparin. Allow 2 h to perform the whole protocol.

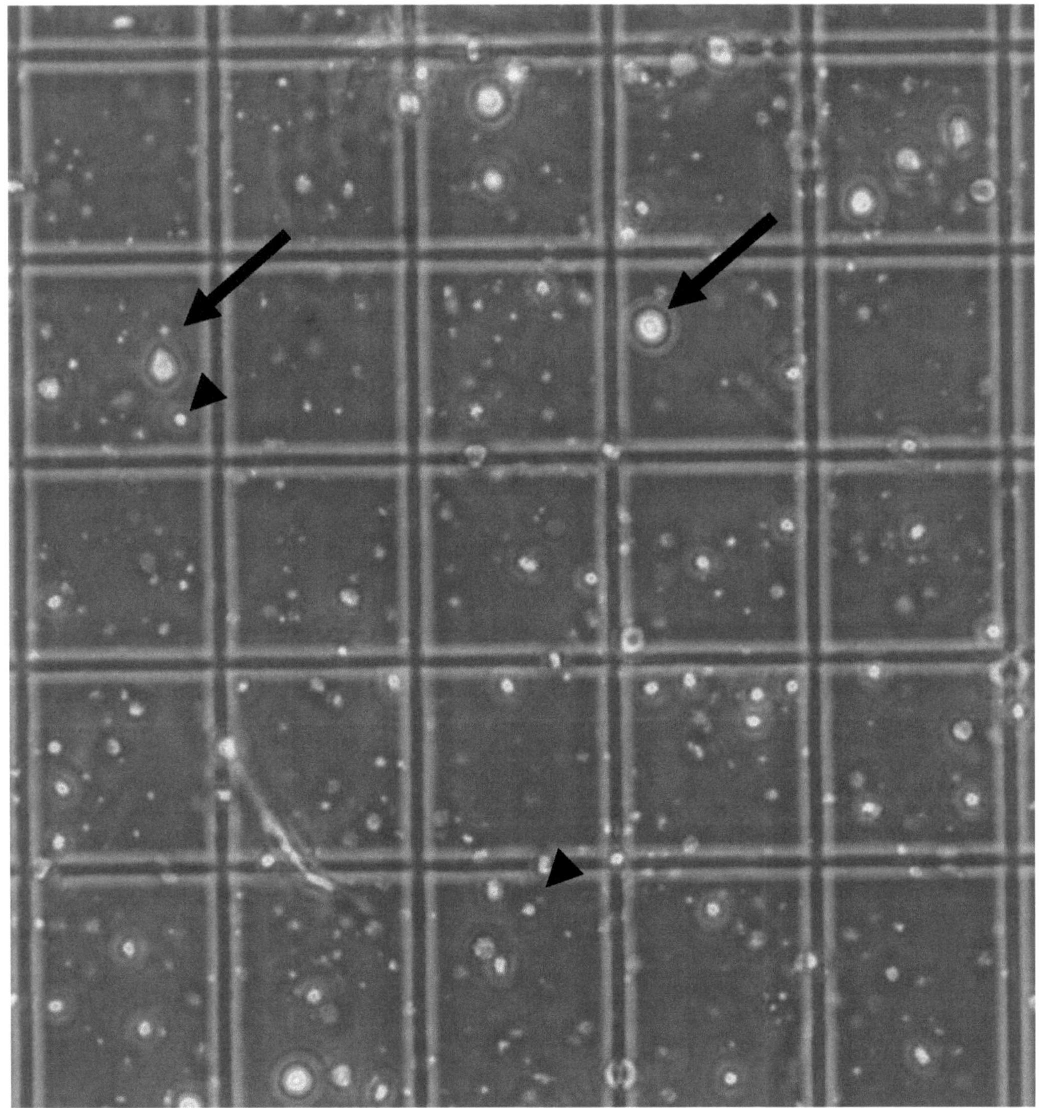

Fig. 2 Aspect of a typical cellular preparation obtained at the end of the protocol (cells were placed in a Malassez counting chamber). *Arrows*: spinal cord cells. *Arrowheads*: red blood cells. Note that due to debris and the various cell sizes, this cell counting is difficult and inaccurate and should be considered as an estimate of the cell number

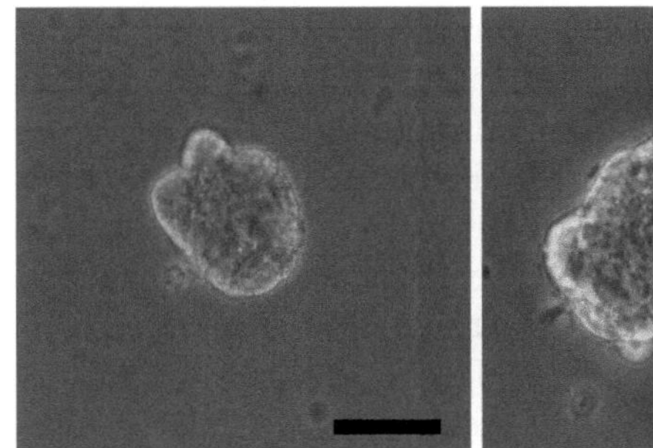
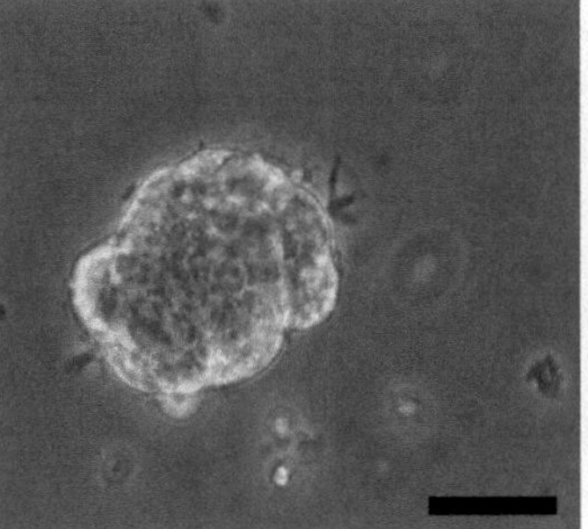
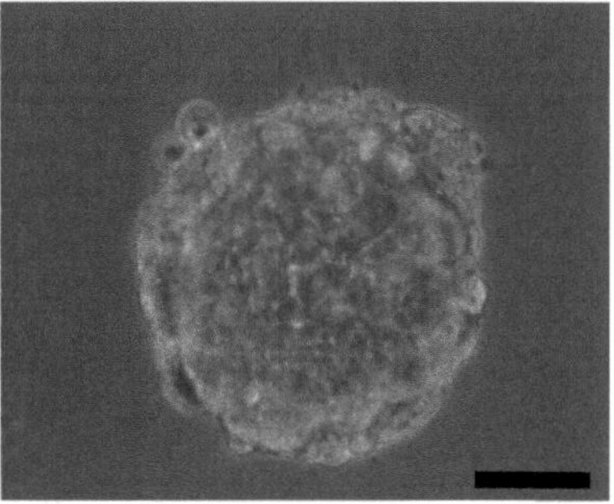

Fig. 3 Examples of neurospheres obtained after 6 days of culture. Note the different neurosphere sizes and the processes sticking out of some cells. Scale bars = 40 μm

3.3 Spinal Cord Neurosphere Culture and Differentiation

1. The emergence of neurospheres should be visible after 5–6 days, first as very refringent small spheres (Fig. 3) where no cells can be distinguished. From this stage on, the spheres grow rapidly and cells become visible. Cellular extensions are often seen at the periphery of the spheres (Fig. 3). These processes are not cilia and move slowly.
2. New FGF2 and EGF are added in the flask every 4 days. When the neurospheres reach a size of approximately 500 μm, these are passaged using enzymatic dissociation as followed.
3. Media is centrifuged for 5 min at 380 × *g* (*see* **Note 6**). After removing the supernatant, the neurospheres are resuspended in 5 ml HBSS w/o Ca^{2+} and Mg^{2+} and centrifuged again.
4. HBSS is removed and the neurospheres are resuspended in 110 μl Trypsin/EDTA solution and incubated for 3 min at 37 °C in a water bath.
5. Add 11 μl of $CaCl_2$ and 2 μl of Dnase I (*see* **Note 9**).
6. Dissociate the neurospheres by gently pipetting up and down with a P200 pipette. After checking under the microscope that the dissociation is correct, 11 μl of trypsin inhibitor is added.
7. The cells are then rinsed with 5–15 ml of HBSS and centrifuge at 380 × *g* (*see* **Note 6**) for 5 min. HBSS is discarded and the cells are resuspended in an appropriate volume of medium.
8. The cells are seeded at a density of 10,000–100,000 cells/ml in a new coated flask (*see* **Note 10**).
9. Proper clonal analysis of neural stem cells is usually performed by seeding one cell per wells in a 96- or 384-well plate. In our experience, after neurosphere dissociation, seeding one cell per well using a cell sorter did not generate new neurospheres unless B27 serum replacement is added in the media.
10. For neurosphere differentiation into neuronal and glial cells, two methods are used. I: Undissociated differentiation is

performed by placing 10–500 neurospheres (diameter typically between 100 and 500 μm, *see* **Note 11**), on a glass coverslip coated with poly-d-lysine/laminin in a well of a 24 wells dish with 0.5–1 ml media without EGF and FGF2. Serum (0.5 %) can be added but to achieve a better astrocytic differentiation. II: Dissociated differentiation is performed by dissociating the neurospheres as described in **steps 3–7** and by seeding 200,000–500,000 cells by wells. This second method allows a better and more accurate quantification of the neuronal and glial populations after performing an immunofluorescence analysis against appropriate cell lineage markers.

4 Notes

1. Do not use 100 % EtOH as the dissolution will take much longer.
2. This step will emit EtOH vapours so it is important to use appropriate dry incubator with non-sparking components to avoid explosion.
3. If the vessels are rinsed with some poly HEMA solution left, this will result in instant formation of white precipitation and the vessel has to be discarded.
4. After death, the mice can be kept at 4 °C for several hours. Even if we have not accurately measured the effect on the yield of neurospheres obtained after an extended period of time after death, stem cells appear to be resistant to the physiological changes that occur after death.
5. It is difficult to dissect the spinal cord without holding it with one hand. We usually spray the gloves with EtOH 70 % to reduce the risk of contamination.
6. 380 and 750 g equals 1,500 and 2,100 rpm, respectively for a rotor with a 15 cm radius.
7. We often use filtered tip at this point to avoid contamination from the pipette tip.
8. It is important to aspirate the layer by approaching slowly the tip of the Pasteur pipette to the top of the layer. If the tip is immersed too deep, the myelin layer will go down and come in contact with the cells which will result in myelin contamination.
9. Dnase I is required at this step as the release of DNA by the cells will result in the formation of big clumps which are very difficult to dissociate. Ca^{2+} is added as Dnase I enzyme needs divalent ions for its activity.
10. In neurosphere culture, it is important to note that the cell growth if very dependent on the initial cell density. High density seeding will generate new spheres within hours as a result of reaggregation. In addition, the sphere released a lot of

endogenous cytokines [13] which help cellular growth. Thus, cultures seeded at relatively high density (for instance 10^6 cells per 75 cm^2 flask) will grow very fast and need to be passaged rapidly (typically within 4–5 days).

11. The number of neurospheres in the flask is estimated by observing 10–20 fields of the flask with a 100× magnification and then by calculating the total neurospheres using the surface observed through the eye piece (it depends on the microscope type but typically around 0.025 cm^2) and the flask surface (25 or 75 cm^2).

Acknowledgments

Our team is supported by the European Union FP6 « Rescue » STREP, the "Association Française contre les Myopathies" (Evry, France) and the "Fondation Thérèse et René Planiol pour l'étude du Cerveau" (Varennes, France).

References

1. Shihabuddin LS et al (2000) Adult spinal cord stem cells generate neurons after transplantation in the adult dentate gyrus. J Neurosci 20: 8727–8735
2. Weiss S et al (1996) Multipotent CNS stem cells are present in the adult mammalian spinal cord and ventricular neuroaxis. J Neurosci 16: 7599–7609
3. Martens DJ, Seaberg RM, van der Kooy D (2002) In vivo infusions of exogenous growth factors into the fourth ventricle of the adult mouse brain increase the proliferation of neural progenitors around the fourth ventricle and the central canal of the spinal cord. Eur J Neurosci 16:1045–1057
4. Meletis K et al (2008) Spinal cord injury reveals multilineage differentiation of ependymal cells. PLoS Biol 6:e182
5. Sabourin JC et al (2009) A mesenchymal-like ZEB1(+) niche harbors dorsal radial glial fibrillary acidic protein-positive stem cells in the spinal cord. Stem Cells 27:2722–2733
6. Horner PJ et al (2000) Proliferation and differentiation of progenitor cells throughout the intact adult rat spinal cord. J Neurosci 20:2218–2228
7. Kulbatski I et al (2007) Oligodendrocytes and radial glia derived from adult rat spinal cord progenitors: morphological and immunocytochemical characterization. J Histochem Cytochem 55:209–222
8. Yamamoto S et al (2001) Transcription factor expression and Notch-dependent regulation of neural progenitors in the adult rat spinal cord. J Neurosci 21:9814–9823
9. Armando S et al (2007) Neurosphere-derived neural cells show region-specific behaviour in vitro. Neuroreport 18:1539–1542
10. Kulbatski I, Tator CH (2009) Region-specific differentiation potential of adult rat spinal cord neural stem/precursors and their plasticity in response to in vitro manipulation. J Histochem Cytochem 57:405–423
11. Hugnot JP, Franzen R (2010) The spinal cord ependymal region: a stem cell niche in the caudal central nervous system. Front Biosci 16: 1044–1059
12. Moreno-Manzano V et al (2009) Activated spinal cord ependymal stem cells rescue neurological function. Stem Cells 27:733–743
13. Deleyrolle L et al (2006) Exogenous and fibroblast growth factor 2/epidermal growth factor-regulated endogenous cytokines regulate neural precursor cell growth and differentiation. Stem Cells 24:748–762

Chapter 6

Culturing and Expansion of "Clinical Grade" Precursors Cells from the Fetal Human Central Nervous System

Maurizio Gelati, Daniela Profico, Massimo Projetti-Pensi, Gianmarco Muzi, Giada Sgaravizzi, and Angelo Luigi Vescovi

Abstract

NSCs have been demonstrated to be very useful in grafts into the mammalian central nervous system to investigate the exploitation of NSC for the therapy of neurodegenerative disorders in animal models of neurodegenerative diseases. To push cell therapy in CNS on stage of clinical application, it is necessary to establish a continuous and standardized, clinical grade (i.e., produced following the good manufacturing practice guidelines) human neural stem cell lines. In this chapter, we illustrate some of the protocols routinely used into our GMP cell bank for the production of "clinical grade" human neural stem cell lines.

Key words Clinical grade, Precursor cells, Human central nervous system, Neural stem cells, Therapy, Good manufacturing practice

1 Introduction

Somatic adult neural stem cells (NSCs) are undifferentiated cells that reside in specialized regions, namely the niche, of the fetal and adult central nervous system (CNS); they possess life-long self-renewal ability and a multipotent differentiation potential, given their ability to generate neurons, astrocytes, and oligodendrocytes. Reynolds and Weiss [1] have first demonstrated a stem cell niche in the CNS. In particular, the finding of adult neurogenesis in the SVZ, which leads to the generation of neural progenitors migrating to the olfactory bulbs and to the cortex, has favored the idea that newborn neurons might subserve cognitive functions and contribute to the homeostasis of the telencephalic–diencephalic area.

During last decades NSCs have been demonstrate to be very useful in grafts into the mammalian CNS to investigate the exploitation of NSC for the therapy of neurodegenerative disorders including both genetic diseases like Metachromatic Leukodystrophy

Brent A. Reynolds and Loic P. Deleyrolle (eds.), *Neural Progenitor Cells: Methods and Protocols*, Methods in Molecular Biology, vol. 1059, DOI 10.1007/978-1-62703-574-3_6, © Springer Science+Business Media New York 2013

(MLD), Huntington's Disease (HD), Alzheimer's Disease (AD) (sporadic) and idiopathic diseases like Parkinson's Disease (PD), Multiple Sclerosis (MS), Amyotrophic Lateral Sclerosis (ALS), stroke [2–5].

Pushed by the encouraging preclinical results into above-mentioned disorders, cell therapy in CNS is reaching the stage of clinical application, with the first clinical trials already underway in some posttraumatic, post-ischemic, tumorigenic, or neurodegenerative disorders (*see* Table 1 for clinical trial ongoing). A continuous and standardized, clinical grade (i.e., produced following the good manufacturing practice guidelines and approved by national and international regulatory agencies) of normal human NSCs would be of paramount importance in regenerative medicine field.

Our adult human hNSCs have now been serially expanded under chemically defined conditions and are being cryopreserved, establishing a Good Manufacturing Practice (GMP)-grade [6], hNSCs bank. These cell bank will be used to perform experimental clinical trials on neurodegenerative diseases. In order to certify these cells by the GMP standard, a panel of cellular, functional and biochemical criteria must be met prior to cell release, which include, but are not limited to, karyotype analysis, stable differentiation and growth capacity, and lack of biological contamination by adventitious agents.

In our GMP facility designed to produce human NSCs for advanced therapies, quality control is only part of overall quality assurance for cell lines which includes evaluation and quality control measures for cells and critical reagents coming into the laboratory, control of the laboratory environment, equipment and procedures, control of data arising from cell culture, control of the delivery of research materials, including cells, to other laboratories, traceability of raw material especially tissue from donors.

Four critical characteristics of cell cultures are fundamental to assure the quality of any cell culture work:

1. Identity, i.e., the cells need to possess a specific behavior:
   (a) Self-Renewal: growth kinetic stable for an elevated number o passages in vitro.
   (b) Multipotency: The cells are able to differentiate into three neural lineages (Astrocytes, Neuron, and oligodendrocytes) after growth factors (EGF and bFGF) removal.
2. Purity, i.e., freedom from microbiological contamination (all the assays need to be performed according to official European Pharmacopeia, current edition [7] in a GMP approved microbiological laboratory) or endotoxins that could prime a dangerous overreaction from patients immune system.
   (a) Sterility (Bacteria and fungi).
   (b) Mycoplasma.
   (c) Bacterial endotoxins.

Table 1
Advanced therapies/regenerative medicine

Cells	Disease	Sponsor	Phase	Title	State
HB1.F3.CD neural stem cells (NSCs Genetically modified)	High-grade Glioma	City of Hope Medical Center	First in man	A Pilot Feasibility Study of Oral 5-Fluorocytosine and Genetically Modified Neural Stem Cells Expressing *E. coli* Cytosine Deaminase for Treatment of Recurrent High-Grade Gliomas	Recruiting
Human spinal cord derived neural stem cell line (Derived from 8-week-old fetus)	Amyotrophic Lateral Sclerosis	Neuralstem, Inc.	Phase I	Human Spinal Cord Derived Neural Stem Cell Transplantation for the Treatment of Amyotrophic Lateral Sclerosis (ALS)	Ongoing
CTX0E03 neural stem cells	Stroke	ReNeuron Limited	Phase I	Pilot Investigation of Stem Cells in Stroke (PISCES)	Recruiting
HuCNS-SC™ (Fetal Neural stem cells)	Neuronal Ceroid Lipofuscinosis	StemCells, Inc.	Phase I	Safety and Efficacy Study of HuCNS-SC in Subjects With Neuronal Ceroid Lipofuscinosis	Terminated
HuCNS-SC™ (Fetal Neural stem cells)	Pelizaeus-Merzbacher Disease	StemCells, Inc.	Phase I	Safety Study of Human CNS Stem Cells Transplantation in Patients With Pelizaeus-Merzbacher Disease (PMD)	Ongoing
HuCNS-SC™ (Fetal Neural stem cells)	Thoracic Spinal Cord Injury	StemCells, Inc.	Phase I/II	Study of Human Central Nervous System Stem Cells (HuCNS-SC) in Patients With Thoracic Spinal Cord Injury	Recruiting
GRNOPC1™ (oligodendrocyte progenitor cells from human embryonic stem cells)	Subacute, Spinal Cord Injury	Geron Corporation	Phase I	Safety Study of GRNOPC1 in Spinal Cord Injury	Recruiting
hNSCs	ALS	Azienda Ospedaliare Santa Maria di Terni	Phase I	Intra-spinal Cord Delivery of Human Neural Stem Cells in ALS Patients: Proposal for a Phase I Study	

Examples of interventional studies, phase I/II, using neural stem cells (source www.clinicaltrials.gov), last 5 years

   (d) Viral contamination.
   (e) BSE (at list risk analysis) contamination.
3. Maintenance of stable functional properties over passaging in vitro.
   (a) Growth curve: constant positive slope over passages.
   (b) Constant ratio of neurons/astrocytes/oligodendrocytes upon differentiation assay.
   (c) Kariology (healthy karyotype asset and deeper analysis like SKY or comparative genomic hybridization).
4. Tumorigenicity, i.e., Cell lines not toxic or tumorigenic.
   (a) Growth factor dependence: the cells died into a few passages after EGF and bFGF removal from culture medium.
   (b) No tumor signs after transplantation into the brain of Nude mice (*see* Fig. 1). The cells are able to migrate, differentiate, and integrate into host tissue.

Because all the characteristics above mentioned, raw materials (media, cell culture plastic disposable, etc.) were obtained from GMP certified suppliers. Human NSCs were produced into controlled environment (class A surrounded by Class B) according to Annex I Vol.4 European GMP Guidelines. Tissue samples were obtained from screened donors according to European Guidelines on "Certain technical requirements for the donation, procurement and testing of human tissues and cell" (Implementing Directive 2004/23/EC of the European Parliament and of the Council); all the procedures where approved by the Ethical Committee (CEAS Umbria).

In this chapter, we illustrate some of the protocols routinely used into our GMP cell bank for the production of "clinical grade" human NSC lines.

2 Materials

2.1 Media

2.1.1 Component List

1. Dulbecco's modified Eagle's medium (DMEM)/F-12 (10×) (Invitrogen).
2. Bovine serum albumin (BSA) (Merck Millipore).
3. 200 mM Glutamine: (100×) (Invitrogen).
4. Heparin pharma grade (FISIOPHARMA, 25,000 UI/5 ml).
5. Glucose (Sigma-Aldrich).
6. HEPES (Invitrogen).
7. $NaHCO_3$ (Invitrogen).
8. Selenium (Sigma-Aldrich).
9. Progesterone (Sigma-Aldrich).

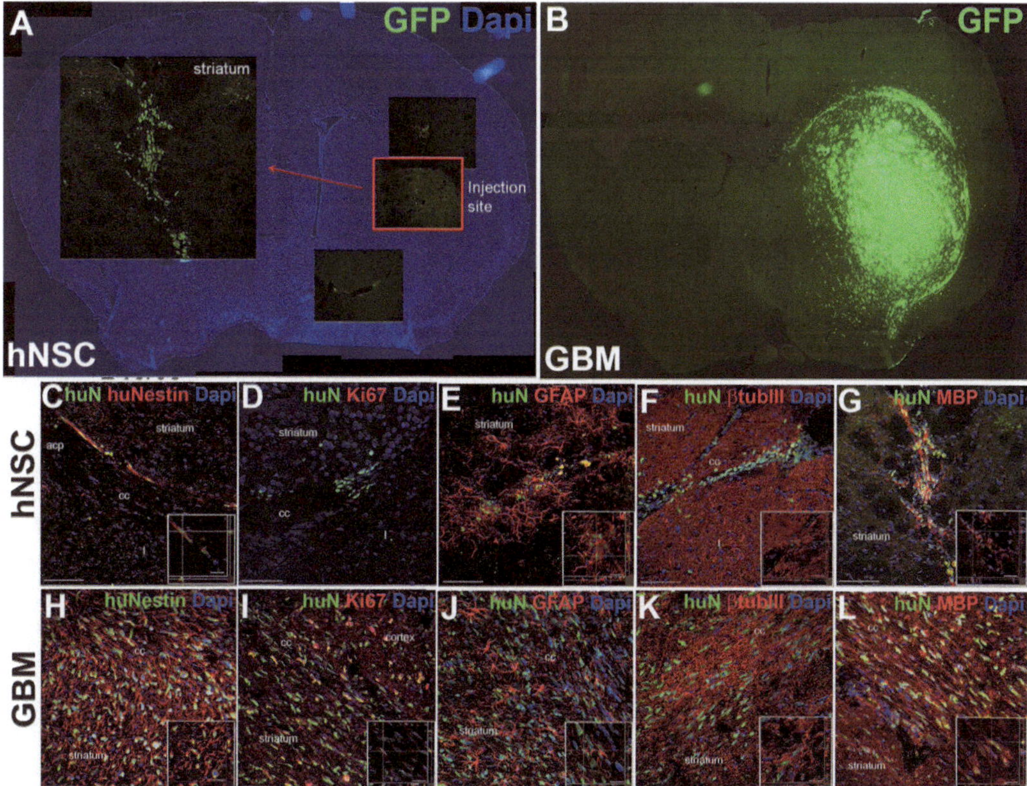

Fig. 1 Nude mice were injected in striatum with hNSCs (**a**) and Glioblastoma cancer stem cells as control (GBM) (**b**). Six months after transplantation mice with hNSCs were sacrificed and immunohistochemistry analysis showed that hNSCs engraft efficiently and migrate throughout the injected hemisphere up to the contralateral (not shown). Mice transplanted with GBM were sacrificed after 4 months and immunohistochemistry analysis showed that cells proliferate extensively and form an invasive tumor cell mass. Confocal microscopy analysis of hNSC (**c**–**g**) and GBM (**h**–**l**) after transplantation into the brain of nude mice. (**a**, **b**) GFP+ (*green*) cells in the striatum and corpus callosum of mice transplanted with hNSC (**a**) and GBM (**b**). (**c**–**h**) huN+ (*green*)/GFAP+ (*red*) hNSC (**c**) and GBM (**h**). (**d**–**i**) huN+ (*green*)/Ki67+ (*red*) hNSC (**d**) and GBM (**i**). (**f**–**k**) huN+ (*green*)/btubulinIII+ (*red*) cells hNSC (**f**) and GBM (**k**). (**e**–**j**) huN+ (*green*)/GFAP+ (*red*) hNSC (**e**) and GBM (**j**). (**g**–**l**) huN+ (*green*)/MBP+ (*red*) hNSC (**g**) and GBM (**l**). Nuclei are shown by dapi staining (*blue*). *Apc* posterior part of anterior commissure, *cc* corpus callosum, *I* intercalated amygdaloid nuclei. Scale bars: in **a**–**j** = 75 μm; insert scale bar: **a**–**i** = 12–18 μm and **f**, **g**, **j** = 20–27 μm. Photo courtesy of Drs. C. Zalfa, D. Ferrari and L. Rota Nodari

10. Putrescine (Sigma-Aldrich).
11. Apo-transferrin (Sigma-Aldrich).
12. Insulin (Sigma-Aldrich).
13. Water for injection.
14. EGF: human recombinant (Peprotech).
15. bFGF: human recombinant (Peprotech).
16. Gentamicin solution pharma grade (FISIOPHARMA, 80 mg/2 ml).

2.1.2 Stock Solutions

1. 3 mM Sodium selenite: Dissolve 100 mg of sodium selenite in 1.93 ml of sterile water (sodium selenite 100×) and store ad −20 °C; Dilute 20 μl of sodium selenite 100× in 1.98 ml of sterile water and store at −20 °C.
2. 2 mM progesterone: Dissolve 100 mg of Progesterone in 15.9 ml of 95 % Ethanol (Progesterone 10×) and store at −20 °C; Dilute 200 μl of 10× Progesterone in 1.8 ml of Ethanol 95 % and store at −20 °C.
3. Recombinant human bFGF: Dissolve 50 μg (1 vial) of bFGF into 500 μl of sterile 5 mM Tris pH 7.6 and Store at −20 °C.
4. Recombinant human EGF: Dissolve 500 μg (1 vial) of EGF into 1,000 μl of sterile water and Store at −20 °C.

2.2 Solutions

1. Phosphate-buffered saline (PBS) 1×.
2. 0.4 % Trypan Blue Solution (SIGMA-ALDRICH).
3. KaryoMax® Colcemid® Solution 10 μg/ml (GIBCO™ Invitrogen Corporation).
4. Acetic acid (Panreac).
5. Methanol (Panreac).
6. Hypotonic solution: 0.075 M Potassium chloride solution (Sigma-Aldrich).
7. CryoSure-DMSO 10 ml EPh/USP grade (LiStar Fish).
8. G banding solution.
9. Giemsa or Leishman stain at 20 % in $Na_2HPO_4 \times 2H_2O$ KH_2PO_4 pH 6.8 buffer solution.
10. 0.9 % NaCl.
11. 1.8 % Trisodium citrate dehydrate.
12. Tissue wash solution: add 12.5 μl of gentamicin to 10 ml of saline solution.
13. Freezing medium solution: 10 % dimethyl sulfoxide (DMSO) in culture media.

2.3 Materials

1. Pasteur glass pipette.
2. Dissecting tools (Martin Instruments, Munich, Germany): scissors, microsurgery clamps.
3. Hemocytometer (Burker chamber).
4. Petri dish 100 mm diameter (Corning).
5. 15 ml conical tube (BD Falcon).
6. Filter tip 1–30 μl (Corning).
7. Filter tip 1–200 μl (Corning).
8. Filter tip 100–1,000 μl (Corning).

9. Flask 25 cm^2, 0.2 μm vented filter cap (Costar Incorporated).
10. Flask 75 cm^2, 0.2 μm vented filter cap (Costar Incorporated).
11. 10 ml sterile plastic pipettes (Corning).
12. 2 ml sterile cryovials (Corning).

2.4 Instruments

1. Programmed freezer down (Planer LNP4-FS).
2. Crio freezer (MVE TEC 3000).
3. Freezing jar (Criostep, Nalgene, PBI International, cat. no. 5100-0001).
4. Water bath or thermo block.
5. Inverted microscope with ocular grid.

3 Methods

The following protocol describes the method to isolate, expand, and characterize hNSC obtained from neural tissue of human fetuses at 8–12 weeks gestation.

3.1 Primary Culture

1. Put the neural tissue into a 100 mm plastic petri dish prefilled with 10 ml of preconditioned medium (at least 1 h at 37 °C, 5 % O_2; 5 % CO_2).
2. Wash the tissue 2 or 3 times in a solution of 0.5 mg/ml gentamicin/physiological solution.
3. Transfer the tissue in a 15 ml tube containing 5 ml of preconditioned medium.
4. Dissociate by triturating using e sterile, fire-polished, cotton-plugged glass pasteur pipette.
5. Let the suspension settle down for 3–4 min.
6. Move the suspension to a clean 15 ml tube, leaving behind the undissociated pieces to the bottom of the tube.
7. Pellet by centrifugation at $110 \times g$ for 10′.
8. Remove the surnatant leaving behind about 160 μl (*see* **Note 1**).
9. Using a 200 μl pipetteman, with the volume set at 120 μl, dissociate the pellet by aspiration/expulsion about 100 times (*see* **Note 2**).
10. Dilute 10 μl of the sample with trypan blue and count in hemocytometer (i.e., Burker chamber) according to European Pharmacopoeia "Nucleated cell count" method, adjusting dilution in order to count at least 100 cells in at least three squares.
11. Seed cells at a density of 1×10^4 cells/cm^2 in preconditioned medium in 25 or 75 cm^2 flask (6 or 12 ml culture media, respectively).

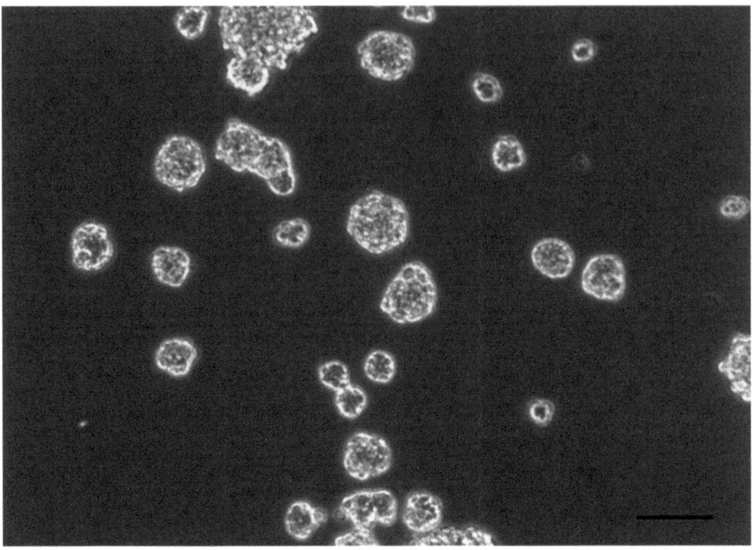

Fig. 2 Human neural stem cells in culture (primary culture). The cells form aggregates named neurospheres. The passages must be made when neurosphere culture is homogeneous and the diameter of the sphere is around 100 μm. 100 μm scale bar

12. Incubate at 37 °C, 5 % O_2; 5 % CO_2 in a humidified incubator (95 % HR) (*see* **Note 3**).
13. Cells should proliferate to form spherical cluster (neurosphere) that should be ready for subculturing (passage) about 7–10 days after plating (*see* Fig. 2).

3.2 Subculturing

1. Transfer the content of the flask in a 15 ml tube.
2. Pellet by centrifugation at 110 × *g* for 10′.
3. Discard the surnatant leaving behind about 160 μl (*see* **Note 1**).
4. Using a 200 μl pipetteman, with the volume set at 120 μl, dissociate the pellet by aspiration/expulsion about 100 times (*see* **Note 2**).
5. Dilute 10 μl of the sample with trypan blue and count in a hemocytometer according to European Pharmacopoeia method.
6. Seed and incubate cells as previously described.

3.3 Cryopreservation

3.3.1 Freezing

(a) Freezing Jar.

This one is the simplest, cheapest, and less time consuming freezing method, already described in a previous work [8].

Briefly:

- After incubation, collect neurosphere and transfer them in a 15 ml tube.
- Pellet by centrifugation at 110 × *g* for 10′.

- Discard all the suspension.
- Add 1 ml of freezing medium (DMSO/culture medium 10 %). Please note that from this point you should be as quick as possible, in order to reduce the possible cellular damage mediated by DMSO.
- Gentle resuspending by pipetting.
- Transfer the cell suspension in a 2 ml cryovial and place the vials into your freezing jar prefilled with isopropyl alcohol.
- Leave the jar at −80 °C for a minimum of 4 h.
- Transfer vial into a liquid nitrogen tank for long-term storage.

(b) Controlled rate slow freezing.

If you need to improve your freezing procedure in terms of less osmotic injury, best vitality after thawing or avoiding intra/extracellular ice formation, the best way is by using a programmable cryogenic freezer, which allows you to apply a slow-controlled freezing rate to your cells. Usually, freezer chamber is connected to a liquid nitrogen supply (i.e., nitrogen tank) by a pump and is sometimes equipped whit a printer which allows you to record the freezing curve (*see* Fig. 3).

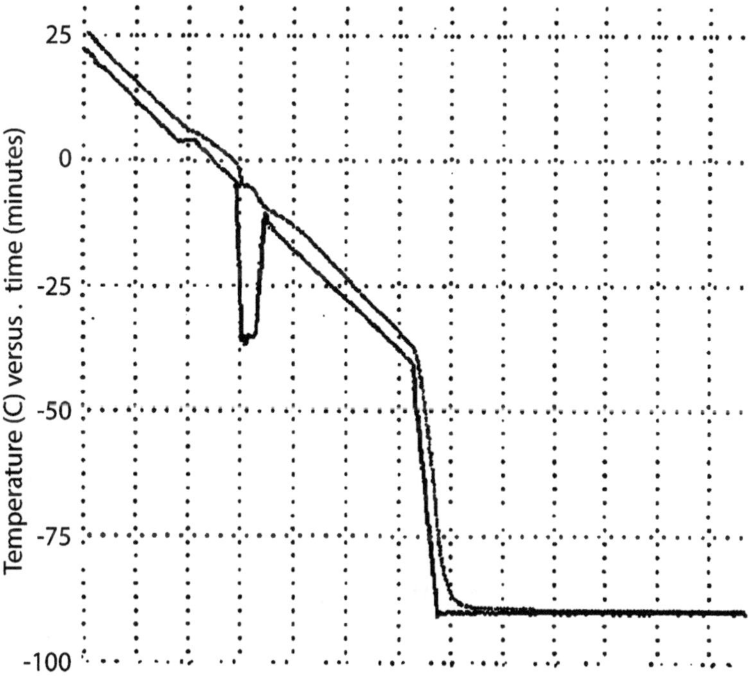

Fig. 3 This figure shown a typical slow controlled freezing curve obtained with Planer instruments

A typical cooling rate of 1 °C/min is appropriate for hNSC, anyway the freezing protocol should be assessed on the basis of the medium components (i.e., fetal serum), cryoprotectant type and dilution and the volume of the sample.

Here we describe a guideline for establishing a freezing protocol for human NSCs. Please note that the freezing protocol performance depends by the volume of the sample, the freezer, the type of vial, the cryoprotectant type and dilution, so you should run some tests to adjust this protocol at your needs.

- After incubation, collect neurosphere and transfer them in a 15 ml tube.
- Pellet by centrifugation at 110 × *g* for 10′.
- Discard all the suspension.
- Add 1 ml of freezing medium (DMSO/culture medium 10 %). Please note that from this point you should be as quick as possible, in order to reduce the possible cellular damage mediated by DMSO.
- Gentle resuspending by pipetting.
- Transfer the cell suspension in a 2 ml cryovial and place the vial into your programmable cryogenic freezer.
- Set the freezing protocol.
  - Start at RT.
  - Decrease temperature to +4 °C at −1 ± 0.5 °C/min.
  - Hold for 5–10 min.
  - Decrease temperature to −4 °C at 1 ± 0.5 °C/min.
  - Decrease temperature to −40 °C at −25 ± 2 °C/min.
  - Hold for 5–10 min.
  - Increase to −15 °C at +10 °C/min.
  - Decrease to −45 °C at 2 ± 0.5 °C/min.
  - Decrease to −90 °C at −15 °C/min.
  - Transfer samples into a liquid/vapor nitrogen tank for long-term storage.

3.3.2 Thawing

- Transfer cryovial(s) from nitrogen to 37 °C water bath or thermo block and leave until thawed.
- Move cellular suspension into 15 ml tube containing 5 ml of preconditioned culture medium.
- Wash by centrifugation at 110 × *g* for 10 min.
- Remove and discard the surnatant.
- Gently transfer pellet in fresh preconditioned medium and seed (*see* **Note 4**).

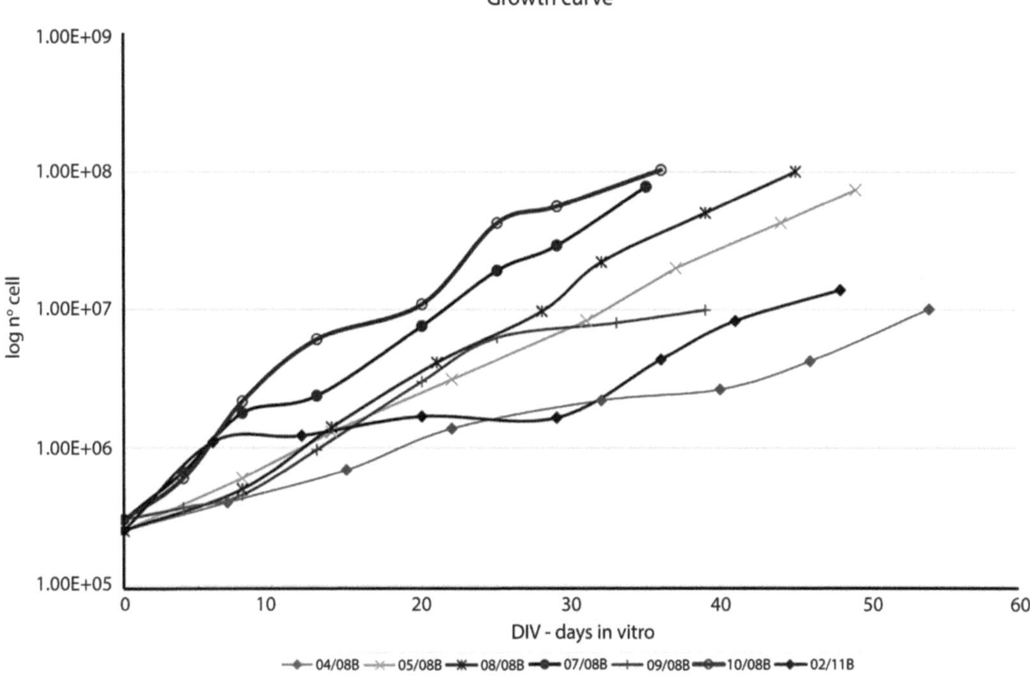

Fig. 4 Kinetic of the growth curve is a direct reflection of self-renewal and it ought to remain stable over time in hNSCs lines

3.4 Characterization

3.4.1 Growth Curve

Construction of a growth curve allows you to know the rate of expansion of your cell line.

- Seed 250,000 cell as previously described in a 25 cm^2 and keep culturing them for at least 5 passages.
- At each passage count the obtained cells and seed again.
- Create a chart representing the logarithmic number of viable cells of each passage, versus the day in vitro since the beginning of the experiment (*see* Fig. 4).

3.4.2 Karyotype Analysis

- Plate 300,000 cells in a 25 cm^2 flask containing 6 ml of preconditioned medium.
- Incubate 37 °C, 5 % O_2; 5 % CO_2 for 48–120 h.
- Add 30 μl of colcemid solution (1:200) and incubate for 3 h.
- Transfer cell suspension in a 15 ml tube.
- Pellet by centrifugation at 110×*g* for 10 min.
- Discard the medium carefully leaving about 200 μl.
- Suspend the pellet using a 200 μl pipetteman.
- Add 10 ml of hypotonic solution and let at 37 °C for 30 min.
- Centrifugate at 110×*g* for 10 min.
- Discard the solution and add the fixative (3:1 methanol:acetic acid), leave 10 min at room temperature.

- Repeat three times the last two operation.
- Proceed with standard Giemsa chromosome staining (if you need you can store the cells suspension in your fridge up to 5 days before staining).
  - Apply 2 or 3 drops of metaphase suspension on your slide.
  - Put the slide over a water bath (37 °C) for 1–2 min.
  - Check under the microscope for the presence of good metaphase spread, if necessary centrifugate again your sample at 110 × *g* for 10 min and discard part of the supernatant, than prepare an other slide.
  - Let the slide dry in incubator overnight.
  - Put in a Trisodium citrate/physiological solution (1:1) for 5 min, than wash in bidistilled water and let dry.
  - Put in a 0.02 % trypsin solution in distilled water for 10–30 s.
  - Wash in physiological solution.
  - Stain with 20 % Giemsa solution for 1–10 min.
  - Wash with water and let dry at room temperature.

4 Notes

1. When the cell pellet is visibly small, less than 160 μl medium should be left in the tube. A too high medium volume does not permit an efficient dissociation of neurospheres.
2. The dissociation of pellets is the crucial step in subculturing stem cells. In order to verify the efficient dissociation of the cells, let a small drop of cell suspension glide down the wall of the tube. If cell clusters are visible, continue pipetting up and down. If neurospheres are not completely disassociated, in subsequent passages cells could form aggregates instead of neurospheres, and finally differentiate in astrocytes, neurons, or oligodendrocytes. However, avoid to pipet more than necessary, as that could induce cell death or cell differentiation, as well.
3. If the cells begin to grow in adherence, the humidity level of the incubator must be checked. A humidity rate too low could induce cells to grow in adherence.
4. 24–48 h after thawing, a further medium renewal is highly recommended, since residual DMSO has toxic effects on stem cells.

References

1. Reynolds BA, Weiss S (1992) Generation of neurons and astrocytes from isolated cells of the adult mammalian central nervous system. Science 255(5052):1707–1710
2. Aboody K et al (2011) Translating stem cell studies to the clinic for CNS repair: current state of the art and the need for a Rosetta Stone. Neuron 70(4):597–613
3. de Filippis L (2011) Neural stem cell-mediated therapy for rare brain diseases: perspectives in the near future for LSDs and MNDs. Histol Histopathol 26(8):1093–1109
4. Kim SU, de Vellis J (2009) Stem cell-based cell therapy in neurological diseases: a review. J Neurosci Res 87(10):2183–2200
5. Lindvall O, Kokaia Z (2010) Stem cells in human neurodegenerative disorders–time for clinical translation? J Clin Invest 120(1): 29–40
6. EU GMP. The rules governing medicinal products in the European Union. Volume 4 EU guidelines to good manufacturing practice—medicinal products for human and veterinary use-, Annex 1–2008
7. European Pharmacopeia 7th edn. volume 2, Chapter 2.6 Biological tests–2010
8. Galli R, Gritti A, Vescovi AL (2008) Adult neural stem cells. Methods Mol Biol 438: 67–84

Chapter 7

Isolating and Culturing of Precursor Cells from the Adult Human Brain

Florian A. Siebzehnrubl and Dennis A. Steindler

Abstract

Adult neural precursor cells are an essential part of the brain, and a focus of two decades of intense research (Ming and Song, Neuron 70:687–702, 2011). Even though adult human stem/progenitor cells have been identified early on (Kirschenbaum et al., Cereb Cortex 4:576–589, 1994; Eriksson et al., Nat Med 4:1313–1317, 1998), progress in the field of adult human neurogenesis has been slow. The reasons for this may be more advanced neighboring fields of pluripotent stem cell research, and lacking study material as well as well-established and standardized protocols. Furthermore, adult precursor cells in humans seem to have greater potential than in rodents (Walton et al., Development 133:3671–3681, 2006). This may be attributed to species differences in astrocyte development and diversity (Oberheim et al., Neurosci 29:3276–3287, 2009).

In this chapter, we provide a guideline for adult human brain tissue dissociation, be it from biopsy or autopsy specimens. This is by no means the only way of culturing adult neural precursor cells, but it may help in streamlining research on this fascinating topic, as well as help introducing others into this field. We describe our methodology for establishing and maintaining long-term cultures from white and grey matter, as well as a simple protocol for differentiating these cells.

Key words Stem cell, Human, Brain, Cell culture, Precursor, Adult

1 Introduction

The last two decades have been an intense period for neurogenesis research, with opinion rapidly changing from (astonished) disbelief to skepticism to wide acceptance [1]. While the number of publications on adult vertebrate neurogenesis now ranges in the thousands, studies on adult human neural stem/progenitor cells and/or neurogenesis are few and far between [2–4]. A number of factors contribute to this phenomenon, from simple lack of donor tissue for research and/or difficulties in tissue acquisition, technical hurdles, time- and material-consuming protocols, to easier-to-procure and faster-to-use human pluripotent stem cells (or their

Brent A. Reynolds and Loic P. Deleyrolle (eds.), *Neural Progenitor Cells: Methods and Protocols*, Methods in Molecular Biology, vol. 1059, DOI 10.1007/978-1-62703-574-3_7, © Springer Science+Business Media New York 2013

progeny). All of these factors have led to an underrepresentation of adult human precursor cells in the current research landscape. Adult neural stem cells have been identified in a fair number of vertebrate species, and neurogenesis is governed by the same principles in all of these species. Nonetheless, adult human neural precursors seem to be a class of their own. This is most notably reflected by the lack of a rodent counterpart. It is currently unclear, whether there are genuine species differences in the potency of human and rodent precursors [5], or whether rodent precursors are so much more potent than human cells that they pass off as stem cells.

In this chapter, we describe a robust method for isolating adult human precursor cells (termed "AHNP" further on for easier readability) from a variety of sources and brain regions. We have tried and tested these protocols in numerous cell culture paradigms, ranging from biopsy to autopsy, healthy to diseased specimens, and 1-year-old to 99-year-old donors. Even though we were successful in most cases, sometimes (and most often unforeseeably) the cultures just would not grow. We advise the cell culturist not to despair but try again. It is our hope that the protocol described here may contribute to advancing research on AHNPs.

2 Materials

2.1 Tissue Preparation

Disclaimer: As human tissue is potentially infectious and often not tested for HIV, hepatitis, or other contagious diseases, protective equipment (gloves, lab coat, goggles) should be worn at all times, and due care should be employed when handling fresh tissue specimens.

2.1.1 Preparation of Solutions and Plates

1. HBSS w/o Calcium and Magnesium (Invitrogen).
2. Papain (Roche).
3. Dispase II (Roche).
4. DNAse I (Roche).
5. $MgSO_4$ (Sigma).
6. 0.22 μm syringe filters (Millipore).
7. 50 cc syringes (BD).
8. Poly-ornithine (Sigma).
9. Laminin (Sigma).
10. Tissue culture plates of various sizes.
11. N2 medium (e.g., Invitrogen).
12. FBS (Atlanta Biologicals).
13. Bovine pituitary extract (Invitrogen).
14. Antibiotics (Sigma).

2.1.2 Tissue Dissociation

1. Sterile 10 cm petri dishes (BD).
2. Sterile scalpel with #11 blade.
3. Sterile scalpel with #21 blade.
4. Sterile forceps.
5. Sterile 50 and 15 ml conical tubes (BD).
6. 70 μm cell strainer (BD).
7. PBS (Invitrogen), sterilized.
8. Hemacytometer.

2.2 Cell Culture

1. Culture medium and coated plates (*see* Subheading 3.1.1).
2. Recombinant human EGF (R&D systems).
3. Recombinant human bFGF (R&D systems).
4. Recombinant human LIF (Millipore).
5. Heparin (Sigma).
6. Accumax (PAA).

2.3 Differentiation Assay

1. Coverslips (coated with Laminin/Poly-Ornithine as described in Subheading 3.1.1).
2. PBS.
3. Paraformaldehyde (Sigma), diluted to 4 % in PBS.
4. Culture medium (*see* Subheading 3.1.1).
5. 3-Isobutyl-1-methylxanthine (IBMX, Sigma).
6. 1-Dibutyryl cyclic AMP (dbcAMP, Sigma).
7. Recombinant human NGF beta (Sigma).

2.4 Immunocytochemistry

1. PBS.
2. Triton-X 100 (Sigma).
3. Blocking solution (5–15 % normal serum in PBS. The blocker varies with the source of antibodies to be used).
4. Forceps.
5. Primary and secondary antibodies.
6. Mounting medium (e.g., Vectashield with DAPI).

3 Methods

3.1 Tissue Preparation

In the following section, we describe a method for establishing cell cultures from human brain tissue specimens. These cultures are a mixed bag of different cell types, but over two or three passages the cultures become homogenous in appearance. Our technique works for both biopsy and autopsy specimens of a large number of brain regions (including cortex, SVZ, hippocampus, and midbrain). We must emphasize the importance of acquiring ethical approval before tissue collection, as well as wearing proper protective equipment when handling tissue specimens. Tissue preparation is performed under a laminar flow hood to reduce the risk of contaminating the resulting cultures. Adding antibiotics to the dissociation solution (Subheading "PPD Solution (Protease–Protease–DNAse)" and Subheading 3.1.2, **step 3**) can further reduce this risk if the tissue specimens are not obtained under sterile conditions (e.g., from autopsies).

3.1.1 Preparation of Solutions

PPD Solution (Protease–Protease– DNAse)

1. Add 100 ml of room-temperature HBSS to autoclaved Erlenmeyer flask.
2. Dissolve 100 mg Dispase II in HBSS.
3. Dissolve 10 mg DNAse I in HBSS.
4. Mix Papain well, add 1 ml to HBSS solution and wait until solution clears.
5. Add 149 mg $MgSO_4$ (12.4 mM, check for hydration status of $MgSO_4$) to HBSS solution, and immediately filter-sterilize with 50 cc syringe and syringe filters.
6. Aliquot into 10 ml portions and keep at −20 °C.

Laminin/Poly-ornithine-Coated Plates or Coverslips

1. Dissolve 150 mg poly-L-ornithine per ml sterile aqua bidest. This stock solution can be stored at −20 °C.
2. Dilute poly-ornithine stock 1:10 in sterile aqua bidest and coat cell culture plates under a laminar flow hood. Incubate over night at room temperature.
3. On the next day, wash plates 3× with sterile aqua bidest. Thaw Laminin on ice and dissolve 15 mg per ml sterile PBS. Add Laminin solution to washed poly-ornithine plates and incubate 2 h at 37 °C.
4. Wash 3× with sterile PBS. Leave last wash of PBS on until use. Laminin/Poly-ornithine-coated plates can be kept at −20 °C.
5. For coating of coverslips, wash glass coverslips in 100 % ethanol, flame off the remaining ethanol and place coverslips into 12- or 24-well culture plates (use glass coverslips of 12 mm diameter for 24-well plates, or 18 mm diameter for 12-well plates). Proceed with coating as above.

Culture Medium

1. Add FBS to N2 medium to a total of 5 %.
2. Add 2.5 μl/ml bovine pituitary extract (BPE) to the medium.
3. Add antibiotics (optional).

Feeding Solution

1. Dilute recombinant human EGF in culture medium to a final concentration of 400 ng/ml.
2. Dilute recombinant human bFGF in the same medium to a final concentration of 400 ng/ml.
3. Dilute recombinant human LIF in the same medium to a final concentration of 200 ng/ml.
4. Add heparin to a final concentration of 2 μg/ml. The complete feeding solution can be stored for 2 weeks at 4 °C.

3.1.2 Dissociation

1. Thaw PPD solution to room temperature. Cut tissue into small fragments (approx. 2–5 mm^3) in sterile petri dish using sterile forceps and sterile #11 blade scalpel (*see* **Note 1**). Mince tissue fragments using sterile #21 scalpel.
2. Collect minced tissue in a 15 ml conical tube by washing the petri dish repeatedly with sterile PBS. The fragments should fit through the bore of a 5 ml pipette.
3. Briefly spin down tissue (1–2 min at 100×*g*), remove supernatant and resuspend pellet in PPD solution (5 ml per 500 mg of tissue). Close the tube tightly and lay down tube against a tube rack to increase the surface. Incubate for 30 min and dissociate tissue every 10 min by pipetting up and down (*see* **Note 2**).
4. Spin down solution for 8 min at 120×*g*, remove supernatant and resuspend pellet in 1 ml PBS. Further dissociate the tissue by pipetting up and down with 1 ml pipette tip. Fill up to 10 ml with PBS, then filter solution through a 70 μm cell strainer, and centrifuge solution again (8 min, 120×*g*).
5. Repeat washing and centrifuging the cell suspension until all or most of the debris has cleared (*see* **Note 3**). Depending on the amount of tissue this should take 2–3 washing steps. Resuspend the final pellet in culture medium.
6. Count the cell suspension on a hemacytometer and plate into appropriate culture dish.

3.2 Tissue Culture

This section describes how to establish and maintain a cell culture from the single cell suspension obtained and plated in the previous section (Fig. 1a).

1. Plate cells at a density of 100,000 cells/ml in culture medium into Laminin/poly-L-ornithine-coated culture dishes. Feed culture with 50 μl/ml feeding solution.
2. Change medium 24–48 h after plating; feed cells.
3. Feed cells every other day, change medium once weekly.
4. When AHNPs have reached confluence (depending on initial number, this may take up to 3 weeks, *see* **Note 4**), split cultures. For this, remove medium and wash once with sterile

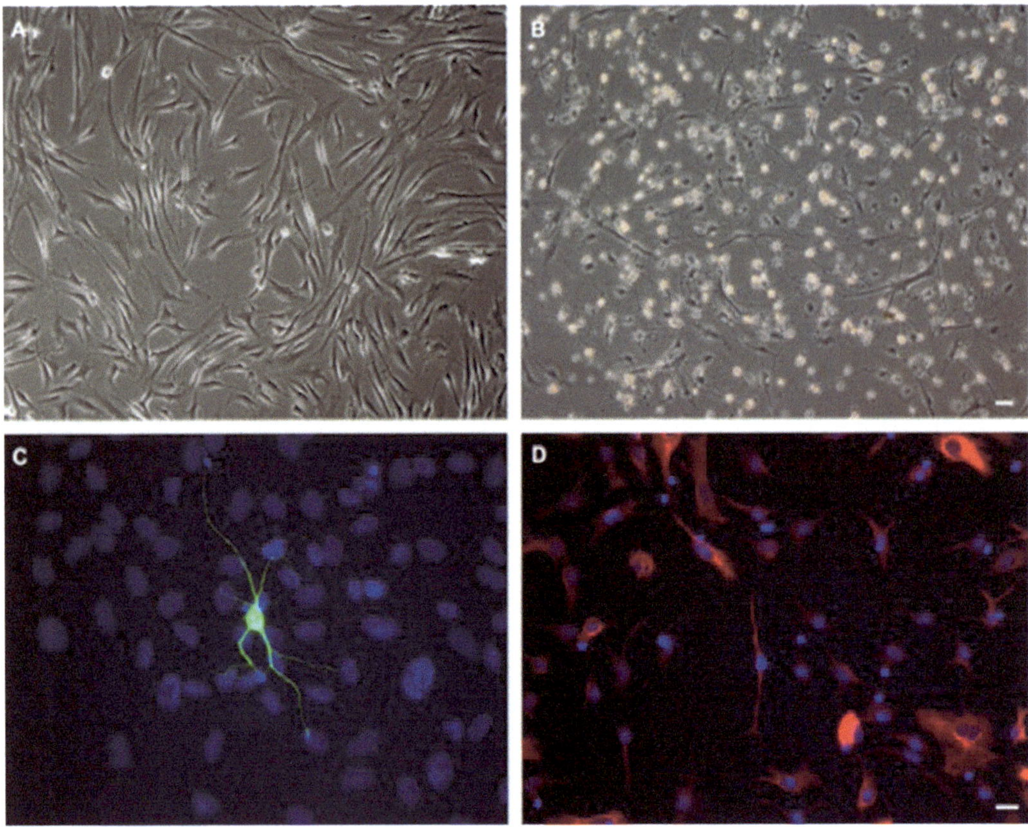

Fig. 1 (**a**) Primitive glial phenotype of AHNPs under expansion culture conditions. This image was taken from a culture in passage 4. (**b**) Differentiation induction generates more complex cellular phenotypes within a short time (7 days). Scale bar (applies also to A) 50 μm. Differentiated AHNPs express neuronal (**c**, bIII tubulin) and glial (**d**, GFAP) markers. Nuclei are counterstained with DAPI. Scale bar (applies to both images) 20 μm

PBS. Add Accumax (half amount of plating medium) to the dish and incubate for 5 min at 37 °C. Collect cell suspension (wash 2–3 times with PBS) and spin down (8 min, 120×*g*). Resuspend cell pellet in plating medium, generating a single cell suspension. Count cells, plate into appropriate dish (*see* **Note 5**) and feed.

3.3 Differentiation Assay

In the final section of this protocol, we will provide a simple cell differentiation assay (Fig. 1b) and the corresponding immunostaining to analyze the content of AHNP cultures.

1. Starting from a confluent AHNP culture, prepare a single cell suspension as outlined in Subheading 3.4.
2. Plate 100,000 cells/ml onto laminin/poly-L-ornithine-coated coverslips in plating medium and feed.
3. Replace medium 24 h after plating with regular culture medium, but do not feed.

4. 48 h after **step 3**, replace culture medium with reduced serum culture medium (i.e., plating medium containing 1 % serum) for control cultures; replace culture medium with reduced serum medium containing 0.5 mM IBMX, 0.5 mM dbcAMP, and 25 ng/ml NGF for neuronal induction cultures.
5. Replace 50 % of culture supernatant with fresh medium daily.
6. 7 days after **step 4**, remove culture medium and replace with 4 % paraformaldehyde in PBS (outside of the laminar flow hood). Incubate for 30 min at room temperature, and then wash three times with PBS.
7. The fixed and washed cells can be used for standard immunochemistry or stored at 4 °C for future use. We recommend wrapping the plates with parafilm for storage.

3.4 Immunocytochemistry

Immunocytochemistry can be used to phenotype the cell types derived from the differentiation assay above (Fig. 1c–d). We use fluorescence-conjugated secondary antibodies to simultaneously detect two antigens, bIII tubulin, and GFAP (*see* **Note 6**).

1. Block cell-containing coverslips with a solution of 10 % FBS in PBS with 0.1 % Triton-X 100. This solution will also be used to dilute antibodies.
2. Incubate coverslips with primary antibodies (mouse anti-bIII tubulin (Promega) 1:1,000 and rabbit anti-GFAP (Dako) 1:1,000) in blocking solution. Incubate for 1 h at room temperature.
3. Wash coverslips three times 5 min each with PBS containing 0.1 % Triton.
4. Incubate coverslips with secondary antibodies (goat anti-mouse Alexa 488 (Invitrogen) 1:500 and goat anti-rabbit Alexa 555 (Invitrogen) 1:500) in blocking solution. Incubate for 30 min at room temperature in the dark.
5. Wash coverslips three times 5 min each with PBS containing 0.1 % Triton.
6. Mount coverslips upside down (i.e., cells pointing downwards) onto slides with Vectashield with DAPI.

4 Notes

1. We collect the tissue into 50 ml conical tubes filled with 25 ml PBS, put the tubes on ice and take the tissue immediately to the lab for preparation. A long forceps can help removing the tissue from the tubes for dissection. We use one set of tools (petri dish, scalpels) per specimen or region of interest to avoid cross-contamination. Adult human brain tissue can be fairly tough, and dissecting and mincing the tissue may take a while.

Again, care should be taken to avoid injuries with the scalpel blade, as human tissue is potentially infectious.

2. The original tissue suspension should fit through a 5 ml pipette tip, and you should be able to switch to a 1 ml pipette tip after the first or second 10 min incubation. The solution should be homogenous at the end of the 30 min dissociation period, but remaining small fragments of white matter are not a problem (these will be removed in during the next step). We usually do not go below 5 ml of preparation solution (although this may be done for very small specimens), and we usually use one tube per 5 ml suspension (as the pipetting and washing steps are easier this way).
3. You should be able to clearly discern the cells and debris in the first pellet. The cell pellet has a more homogenous and compact appearance and is below the debris (and usually much smaller). Depending on the amount of starting tissue it may be helpful to distribute the cell suspension into two or three tubes of 10–15 ml, otherwise it may take a long time until the debris is cleared away. The tissue suspension should not be centrifuged in PBS for extended amounts of time, as cell viability may be affected.
4. When we speak of confluence, we mean about 80 % confluence. AHNP cultures should be split before they grow completely confluent, otherwise the cells may enter senescence.
5. Usually, cells are split 1:2 or 1:3 for expansion. We have been able to culture adult human hippocampal precursor cells for extended time periods using this protocol.
6. The amount of neuronal differentiation in unstimulated culture conditions is fairly low (usually around 1 %), but can be increased by stimulating AHNPs with IBMX, dbcAMP, and NGF. This should lead to a dramatic increase in the number of neuronal phenotypes.

References

1. Ming GL, Song H (2011) Adult neurogenesis in the mammalian brain: significant answers and significant questions. Neuron 70:687–702
2. Kirschenbaum B et al (1994) In vitro neuronal production and differentiation by precursor cells derived from the adult human forebrain. Cereb Cortex 4:576–589
3. Eriksson PS et al (1998) Neurogenesis in the adult human hippocampus. Nat Med 4:1313–1317
4. Walton NM et al (2006) Derivation and large-scale expansion of multipotent astroglial neural progenitors from adult human brain. Development 133:3671–3681
5. Oberheim NA et al (2009) Uniquely hominid features of adult human astrocytes. J Neurosci 29:3276–3287

Chapter 8

Isolation and Culture of Precursor Cells from the Adult Human Spinal Cord

Luc Bauchet, Nicolas Lonjon, Florence Vachiery-Lahaye, Alain Boularan, Alain Privat, and Jean-Philippe Hugnot

Abstract

Our group recently provided evidence for the presence of neural stem cells and/or progenitor cells in the adult human spinal cord. In this chapter, we review materials and methods to harvest high-quality samples of thoracolumbar, lumbar, and sacral adult human spinal cord from brain-dead patients who had agreed to donate their bodies to science for therapeutic and scientific advances. The methods to culture precursor cells from the adult human spinal cord are also described.

Key words Adult human spinal cord, Stem cells, Precursor cells, Organ donation

1 Introduction

The brain of adult humans and rodents contain neural stem and progenitor cells; furthermore the presence of neural stem cells and progenitor cells in the adult rodent spinal cord has been previously described. In our first publication [1], using electron microscopy, expression of neural precursor cell markers, and cell culture, we investigated whether neural precursor cells were also present in the adult human spinal cord. In well-preserved non-pathological post-mortem samples of human adult spinal cord, we found that nestin, Sox2, GFAP, CD15, Nkx6.1, and PSA-NCAM were expressed heterogeneously by the cells located around the central canal. Ultrastructural analysis revealed the existence of immature cells close to ependymal cells, which displayed characteristics of type B and C cells found in the subventricular region of the adult rodent brain, respectively, considered to be stem and progenitor cells. Completely dissociated spinal cord cells formed, in a reproducible manner, Sox2+ nestin+ neurospheres containing proliferative precursor cells. On differentiation, these generated glial cells and g-aminobutyric acid (GABA)-ergic neurons. These results provided

Brent A. Reynolds and Loic P. Deleyrolle (eds.), *Neural Progenitor Cells: Methods and Protocols*, Methods in Molecular Biology, vol. 1059, DOI 10.1007/978-1-62703-574-3_8, © Springer Science+Business Media New York 2013

the first evidence of the existence in the adult human spinal cord of neural precursors with the potential to differentiate into neurons and glia. These results are highly relevant not only for the endogenous regeneration of spinal cord after trauma but also for degenerative diseases affecting the spinal cord.

More recently, we also reported that smooth muscle cells expressing Nestin and Nkx6.1 are one of the main cell populations derived from culturing human spinal cord cells in adherent conditions with serum and might be involved in scar formation during spinal cord injury [2]. Moreover, the high quality of the available adult human spinal cord samples allowed us to participate in an European Program (STREP RESCUE), where we described the anatomy of serotoninergic projections in human spinal cord at thoracic and lumbar levels [3].

In this study, we review in details the materials and methods necessary for getting high-quality samples of thoracolumbar, lumbar, and sacral adult human spinal cord from brain-dead patients who had agreed to donate their bodies to science for therapeutic and scientific advances. The methods to culture precursor cells from the adult human spinal cord are also described.

2 Materials

2.1 Patient Conditions and Materials for Organ Harvesting (in a Context of Organ Transplantation)

2.1.1. Operating room fitted to carry-out organ harvesting for therapeutic and scientific purposes.

2.1.2. Surgical and anesthetic teams and materials to perform organ transplantation.

2.1.3. Brain-dead patient (with a beating heart), meeting all criteria (e.g., clinical, biological, legal, and ethical) for organ donation, according to the established guidelines by the French Biomedical Agency [4, 5]. (Guidelines established in France are presented in **Notes 1** and **2**).

2.1.4. Organ preservation solution (IGL-1, Institute Georges Lopez, France).

2.1.5. Sterile ice, chilled sterile drapes (they will be positioned into the abdomen when proceeding with organ harvesting).

2.2 Neurosurgical Team and Specific Materials

2.2.1. One neurosurgical team involved in spinal cord and spine surgery (*see* **Note 3**).

2.2.2. Surgical instruments for spinal cord surgery: bipolar forceps, mallet, chisels, needles holder, raspatorium, periosteal elevators, surgical rongeurs, bone rongeurs, bone curettes, osteotomes, and surgical gouges; oscillating saw and hand pieces; anterior self-retaining retractor system for abdominal surgery.

2.2.3. Delicate surgical instruments: delicate and dissecting scissors: Metzenbaum and Mayo; delicate dissecting forceps: Adson, Debakey; scalpel blades and scalpel handles; ruler.

2.2.4. Posterior instrumentation device: posterior spinal instrumentation and ancillary for the specific spinal osteosynthesis system; used rod length between 30 and 50 cm, pedicle screw length between 35 and 45 mm.

2.2.5. Instruments for abdominal surgery: universal needle holder and surgical thread.

2.2.6. The materials, solutions, and culture medium used to prepare and culture human spinal cord stem cells are the same as those described for the mouse spinal cord stem cell culture (Chapter 5).

3 Methods

Here, the human spinal cord samples come from brain-dead patients who had agreed to be donors for organ transplantation. All guidelines used for organ donations have been applied here [5]. Technical recommendations for organ and tissue harvesting in brain-dead donors were also published by the French Biomedical Agency [6] (*see* **Notes 1** and **2**). We detail here the neurosurgical procedure required to get spinal cord samples and the methods to culture precursor cells from adult human spinal cord.

3.1 Exposure and Removal of the Thoracolumbar Spine

The anterior exposure to the thoracolumbar spine (T8/T9-L2/L3) is performed after having harvested the organs for transplantation, with the same approach. The transperitoneal associated with the transsternal approach provides a good visualization of the lumbar and thoracic spine. The more internal organs (kidneys, liver, heart, and lungs) were taken out, the better the exposure to the vertebral column was thus simplifying the procedure. Abdominal viscera can easily be retracted away from the spine.

The iliac vessel and L5-S1 disc can clearly be seen and above it we identify and section with a knife blade the L2-L3 disc. In this space, going deeper, we make an incision into the thecal sac and we cut entirely the cauda equina.

After proper exposure, the transverse processes are prepared using a Bovie electrocautery unit and a Cobb periosteal elevator or raspatorium. A subperiosteal resection of the psoas muscle allows for proper exposure of the rostral lumbar spine. Then dividing the diaphragmatic crus from the anterior longitudinal ligament and subsequently from the arcuate ligament in the transverse process of

L1 facilitates the exposure of the caudal extent of D12. Taking off the heart-lung block allows for proper exposure of the middle and lower thoracic spine. The disc L2-L3 is marked and we can then easily count up to the eighth thoracic vertebra.

The disc surrounding the T8 or T9 level is resected using a knife or osteotome and removed by an anterior approach using pituitary rongeurs. The posterior longitudinal ligament followed by the thecal sac with the spinal cord is also resected. Between L3 and T8, at each level and on each side a costotransversectomy is done and the transverse process is excised with an oscillating saw blade or manual osteotome. After the transverse section on the lumbar spine and the costotransversectomy on the thoracic spine we can elevate the vertebral block between L2 and T8. We do an en bloc resection of the vertebras after using scissors to cut the muscle and aponeurotic attachments to the spinous process (*see* **Note 4**).

Then, one team dissects the spinal cord from the vertebral bloc and prepares the spinal cord sample for stem cell culture, while the other team performs the osteosynthesis and surgical closure.

3.2 Dissection of the Thoracolumbar, Lumbar, and Sacral Spinal Cord (Fig. 1a)

On another operating table, the vertebral bloc formed by about six vertebrae (between five or seven vertebrae) is dissected. After removing the posterior arch of each vertebra (either by splitting the pedicles at vertebral levels or by opening the spinal canal with a multilevel laminectomy), the spinal cord with the proximal cauda equina is removed along with the spinal meninges (dura mater, arachnoid, and pia mater). Then, the spinal cord and the dorsal root ganglia are resected and separated from the meninges (*see* **Notes 5** and **6**).

3.3 Preparation for Human Spinal Cord Stem Cell Culture

3.3.1. The spinal cord (Fig. 1a) is cut into 0.5–1 cm thick fragments with a scalpel.

3.3.2. The fragment to be used for cell culture is placed into a Petri dish with a small volume of PBS and the remaining meninges are removed with forceps as much as possible (Fig. 1b) (Note: this can also be done before cutting the spinal cord into fragments).

3.3.3. Each fragment is weighed by placing it in sterile preweighed tube.

3.3.4. Mince the fragments with fine scissors by placing them into a Petri dish lid.

3.3.5. Transfer the tissue with a spatula into a 15-ml tube and add HBSS with Ca^{2+} and Mg^{2+} to a final volume of 11 ml.

3.3.6. For a tissue weighing 2 g, add 1.3 ml Hyaluronydase, 0.3 ml Kynureic Acid, 1.3 ml Trypsine, 0.1 ml DNAse I, and incubate for 30–50 min at 37 °C in water bath. Every 5–10 min, the suspension is homogenized by pipetting up and down with a 10 ml pipette.

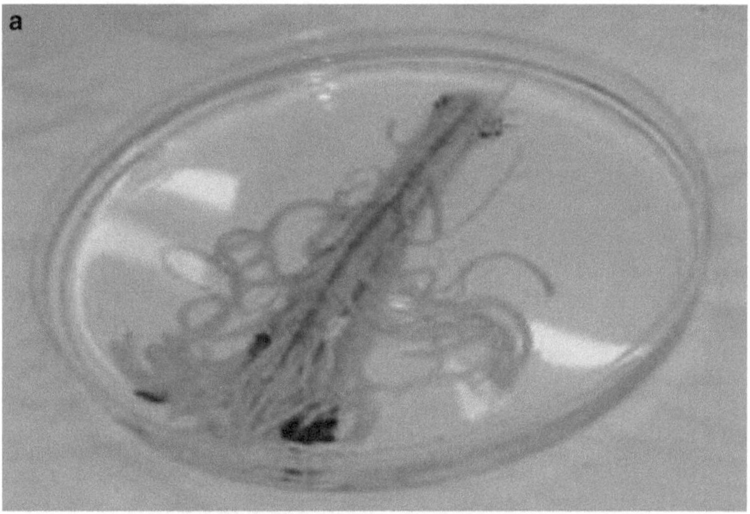

Fig. 1 (**a**) Aspect of a human spinal cord. (**b**) Aspect of a fragment from which meninges have been removed. Scale bar = 1 cm

3.3.7. Centrifuge at 380 × *g* for 5 min.

3.3.8. Remove the supernatant and resuspend the pellet with 5 ml of HBSS w/o Ca^{2+} and Mg^{2+}.

3.3.9. Neutralize the remaining trypsin by adding 0.3 ml of the 5 % trypsin inhibitor solution.

3.3.10. Dissociate the spinal cord fragment by gently pipetting up and down with a P1000 pipetman fitted with a blue tip.

3.3.11. Fit a 100 μm cell strainer (yellow) onto a 50 ml centrifuge tube and transfer the cellular suspension into the filter, let it go through and rinse with a 1.7 ml HBSS w/o Ca^{2+} and Mg^{2+}.

3.3.12. Transfer to a 15 ml tube and centrifuge at 380 × *g* for 5 min. Remove the supernatant and resuspend in 25–50 ml of sucrose solution.

3.3.13. Centrifuge at 750 × *g* for 30–40 min. Very carefully, remove the tubes from rotor and place them on a rack, avoiding any abrupt movements.

3.3.14. Eliminate the top white layer (myelin) and then the entire supernatant by aspiration using a Pasteur pipette or a yellow tip connected to a vacuum pump.

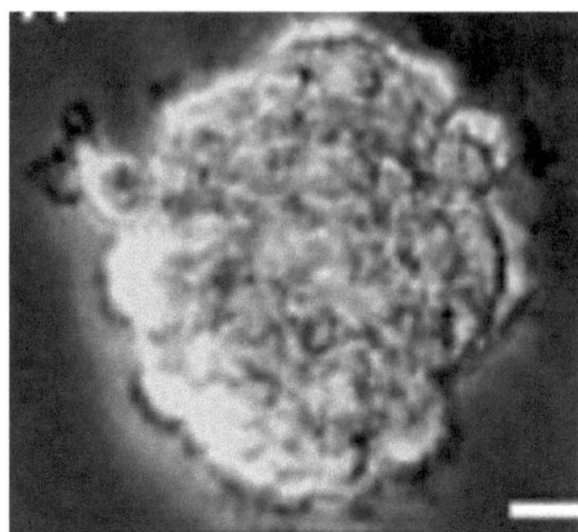

Fig. 2 Neurosphere culture of adult human spinal cord. Scale bar = 20 μm

3.3.15. Resuspend the pellet in 1 ml of complete medium and perform a cell count. The yield is typically between 2 and 4×10^6 cells per g of tissue.

3.3.16. Cells are seeded in poly-HEMA-coated flasks at a density of 1–200,000 cells/ml with EGF, FGF2, and heparin as described for the mouse spinal cord stem cells. Neurospheres are typically observed after 2–4 weeks (Fig. 2).

3.4 Osteosynthesis

After multiple vertebrectomies, anterior spinal fixation is done with a posterior instrumentation device. Pedicle screws are placed in the vertebra corpus. The choice of the screw length and diameter and rod length is based on the knowledge of the patient's anatomy. Usually, we use two rods and two or four pedicle screws fixed one or two levels just above and below the resection area.

3.5 Surgical Closure

We perform a conventional abdominal wall and sternum closure using commonly used surgical thread.

4 Notes

1. Here, human spinal cord samples come from brain-dead patients, who were organ donors. Brain death usually occurs suddenly in cases of severe brain damage secondary to head trauma, lack of oxygen to the brain or stroke in patients following in intensive care units. Organ transplantation procedure is performed in strict accordance to the French laws. Brain death is the irreversible end of all brain activity (including involuntary activity necessary to sustain life). A brain-dead person has no clinical evidence of brain function during a physical examination. This includes no response to pain, no cranial nerve reflexes and no spontaneous breathing. The diagnosis of brain death needs to be extremely in order to validate that the condition is irreversible.

The patient should have a normal temperature and be drug-free since drugs can suppress brain activity if the diagnosis is to be based on an electroencephalogram (EEG). Two EEGs (respecting at least a 4-h interval) should be flat, or a cerebral blood flow CT-scan must validate the complete absence of intracranial blood flow. Because we are part of the organ donation process, blood group, HLA typing, viral serology, and search for infectious diseases or neoplasia are always performed in our unit.

2. The integrity of the human body is fully respected. No further skin incision is made. The surgical approach used is that of organ harvesting for therapeutic purposes. Spinal stability is restored and we use human spine osteosynthesis instruments. Surgical closure is exactly the same as that used for the removal of organs for therapeutic purposes.
3. At least one senior neurosurgeon and one junior neurosurgeon are needed. This being the minimum required. An additional assistant is very helpful to help the senior neurosurgeon remove the spinal cord from the spine, while the second assistant neurosurgeon starts the osteosynthesis and the surgical closure.
4. For the first samples taken, we initially performed a slightly different method. Instead of taking a piece of the spine from L2 to T9 with the en bloc approach, we made successive corpectomies and discectomies from L2 to T9, and we then removed the spinal cord and the beginning of the cauda equina. This first method is actually longer and it can lead to additional risks of injuring the spinal cord.
5. The duration of vascular clamping was an average of 4 h, when heart, lungs, liver, and kidneys were removed, and an average of 2 h when only the liver and kidneys were removed. After harvesting the organs for therapeutic outcomes, the duration of the spinal cord removal was an average of 1.5 h.
6. All procedures are performed in sterile conditions.

References

1. Dromard C et al (2008) Adult human spinal cord harbors neural precursor cells that generate neurons and glial cells in vitro. J Neurosci Res 86:1916–1926
2. Mamaeva D et al. (2011) Isolation of mineralizing Nestin+ Nkx6.1+ vascular muscular cells from the adult human spinal cord. BMC Neurosci Oct 10;12:99
3. Perrin FE et al. (2011) Anatomical study of serotonergic innervation and 5-HT(1A) receptor in the human spinal cord. Cell Death Dis Oct 13;2:e218
4. Camby C (2008) The French Biomedicine Agency and medically assisted reproduction. Bull Acad Natl Med 192:17–21
5. French Biomedicine Agency (2010) http://www.agence-biomedecine.fr/uploads/document/doc_agenceva.pdf
6. Tixier D, Barrou B (2007) Recommandations techniques pour le prélèvement des organes et des tissus sur donneurs en état de mort encéphalique. http://www.edimark.fr/publications/articles/prelevement-des-organes-et-des-tissus-sur-donneurs-en-etat-de-mort-encephalique/13358

Chapter 9

Isolation and Enrichment of Defined Neural Cell Populations from Heterogeneous Neural Stem Cell Progeny

Hassan Azari

Abstract

The renewable source of neural stem cells (NSCs) with multi-lineage differentiation capability towards neurons, astrocytes, and oligodendrocytes represent an ideal supply for cell therapy of central nervous system (CNS) diseases. In spite of this, the clinical use of NSCs is hampered by heterogeneity, poor neuronal cell yield, predominant astrocytic differentiation of NSC progeny and possible uncontrolled proliferation, and tumor formation upon transplantation. The ability to generate highly enriched and defined neural cell populations from the renewable source of NSCs might overcome many of these impediments and pave the way towards their successful clinical applications.

Here, we describe a simple method for NSC differentiation and subsequent purification of neuronal progenitor cells, taking advantage of size and granularity differences between neuronal cells and other NSC progeny. This highly enriched neuronal cell population provides an invaluable source of cells for both in vitro and in vivo studies.

Key words Neural stem cell, Neuronal progenitor cells, Flow cytometry, Isolation, Enrichment

1 Introduction

Neural stem cells (NSCs) were first isolated and expanded from the adult mammalian brain in 1992 using a serum-free culture system referred to as the neurosphere assay [1–4]. This renewable source of multipotential stem cells can give rise to all three major neural cell types of the central nervous system (CNS) [5], which makes it a promising candidate for cell replacement therapy in different neurological diseases. Despite their un-limited proliferative potential and self-renewal properties, NSC progeny are very heterogeneous (i.e., including *bona fide* NSC, lineage-restricted neuronal and glial progenitors, and differentiated cells) [6, 7]. Furthermore, in contrast to their in vivo counterparts in the subventricular zone (SVZ) and the subgranular zone (SGZ) of the brain, where active neurogenesis is occurring [8], the in vitro expanded NSCs are less

Brent A. Reynolds and Loic P. Deleyrolle (eds.), *Neural Progenitor Cells: Methods and Protocols*, Methods in Molecular Biology, vol. 1059, DOI 10.1007/978-1-62703-574-3_9, © Springer Science+Business Media New York 2013

neurogenic and are mainly biased towards an astrocytic fate upon differentiation both in vitro and upon transplantation [9, 10]. The lack of control over their fate upon transplantation into a diseased CNS environment, especially to a non-neurogenic region, and also the possible uncontrolled proliferation and tumor formation of implanted NSCs [11] are the two main impediments towards their practical application in a clinical setting. Using defined neural cell populations (i.e., neuronal, astrocytic and oligodendroglial progenitor cells) with a lineage-restricted fate would greatly resolve these limitations and provides us with the ability to select and implant the cells of interest at a predefined ratio and dosage depending on the nature and/or stage of the disease.

Flow cytometry technology provides a powerful means for the isolation and enrichment of many different cell populations based on their size, granularity, and antigens expressed on the cell surface [12–16]. Although proliferating neural stem/progenitor cells harvested from different parts of the CNS do not differ significantly in their size and granularity, differentiating NSC progeny show a sharp difference both in size and granularity that can be exploited by flow cytometry, where distinct cell populations can be isolated based on light scatter properties [15, 16].

In this chapter, we describe a simple two-step differentiation method to generate neuronal progenitor cells from NSCs harvested from the ganglionic eminences of embryonic day-14 mouse brains using the neurosphere assay. Following differentiation, the neuronal progenitor cells are purified based on their unique size and granularity (low Forward Scatter (FSC^{low}) and low Side Scatter (SSC^{low}) properties) in comparison to other NSC progeny using flow cytometry. Alternatively, a higher neuronal purity is achieved by positively selecting the PSA-NCAM-immunoreactive (IR) neuronal progenitors within the $FSC^{low}SSC^{low}$ cell population. Finally, an immunofluorescent method will be used to discriminate the phenotypes of the sorted cells.

2 Materials

2.1 Cell Culture

2.1.1 Cell Culture Plasticware

1. Flasks: 25, 75, and 175 cm^2 with 0.2 μm vented filter cap.
2. Sterile 15, 50 ml polypropylene conical tubes.
3. 40 μm cell strainer.
4. 0.22-mm bottle-top filter.
5. Tissue culture treated 96-well plates.
6. Media bottles (100, 250, or 500 ml).

2.1.2 Culture Medium, Supplements, and Reagents

1. Basal medium (NeuroCult® NSC basal medium, StemCell Technologies).

2. Proliferation supplement (NeuroCult® NSC Proliferation Supplements, StemCell Technologies).
3. Fetal calf serum (FCS).
4. 0.05 % Trypsin/EDTA.
5. Phosphate-buffered saline.
6. Minimal essential medium (MEM).
7. HEPES-buffered minimum essential medium (HEM): Mix 1 × 10 l packet of MEM and 160 ml of 1 M HEPES and bring the volume to 8.75 l using distilled water. Set the final PH to 7.4 and store it at 4 °C.
8. Trypsin inhibitor solution: First make 10 ml of DNase solution (100 mg DNase dissolved in 100 ml of HEM) and then add 0.14 g of trypsin inhibitor to DNase solution and finally make the volume up to 1 l using HEM. Keep aliquots of the final products in −20 °C freezer.
9. Stock solution of epidermal growth factor (EGF, 10 μg/ml).
10. Stock solution of basic fibroblast growth factor (b-FGF, 10 μg/ml).
11. Stock solution of heparin (of 0.2 % in PBS).
12. Poly-L-Ornithine.
13. Human recombinant BMP-4.

2.2 Immunofluorescent Staining

1. 4 % Paraformaldehyde (PFA) in PBS; pH 7.4.
2. Triton-X 100 to be used in primary and secondary antibody solutions.
3. Primary and secondary antibody solution: a 0.1 % PBS-Triton-X 100 + 5 % Normal Goat serum (NGS) solution is prepared and then sufficient amount of antibodies (*see* Table 1) is added to the solution.

2.3 Flow Cytometry

1. PBS.
2. Mouse Anti-PSA-NCAM monoclonal antibody, conjugated to Phycoerythrin (PE).
3. PE-conjugated mouse IgG1 isotype control antibody.
4. Fluorescence-activated Cell Sorter.
5. Propidium iodide solution; dissolve PI in sterile water to a final concentration of 500 μg/ml.

3 Methods

In the following section, first the differentiation and then the flow cytometry-based isolation and purification of neuronal progenitor cells from the differentiating NSC progeny will be described.

Table 1
Primary antibodies and targeted antigens for the different neural cell lineages

	Antigen	Working dilution	Source
Neurons	βIII-tubulin	1:2,000	Promega#G7121
	Microtubule-associated protein-2 (MAP-2)	1:300	Chemicon # MAB3418
	Double cortin	1:1,000	Chemicon # AB5910
	PSA-NCAM	1:300	Chemicon # MAB5324
	DARPP-32	1:500	Sigma#AB4503329
	GAD-65/67	1:500	Millipore#AB1511
Astrocytes	Glial fibrillary acidic protein (GFAP)	1:500	Dako Cytomation # Z0334
Oligodendrocytes	O4	1:300	Chemicon # MAB345
	Gal-c	1: 300	Chemicon # MAB342

To achieve consistent results, using low-passaged NSCs (neurospheres) is highly recommended.

3.1 Neural Stem Cell Preparation

1. Isolate and expand neural stem and progenitor cells from adult SVZ or embryonic day 14 ganglionic eminences as described in separate protocols [1–4].
2. Collect and transfer the neurospheres (150–200 μm in diameter) from tissue culture flask(s) to a right size sterile tissue culture tube, and pellet the spheres at 800 rpm (110×*g*) for 5 min at room temperature (*see* **Note 1**).
3. Discard the supernatant, resuspend the pellet in 1 ml of prewarmed 0.05 % trypsin–EDTA and incubate the spheres in a 37 °C water bath for 2–3 min.
4. Mix the neurosphere suspension with an equal volume of soybean trypsin inhibitor to stop the trypsin activity (*see* **Note 2**).
5. Gently pipette up and down the resulting suspension for 3–5 times to ensure complete trypsin inactivation and neurosphere dissociation (*see* **Note 3**).
6. Again pellet the cell suspension at 800 rpm (110×*g*) for 5 min, discard the supernatant and resuspend the pellet in 1 ml of complete NSC medium.
7. Add 10 μl of the cell suspension to 90 μl of Trypan blue (0.04 % in PBS). Mix well and take 10 μl of the resulting suspension to perform a cell count using a hemocytometer.

3.2 Differentiation of Neural Stem Cells Using the Neuroblast Assay (NBA) Method

The neuroblast assay differentiation method has two stages: *proliferation stage* in which neural stem and progenitor cells actively proliferate and generate a cell monolayer and *differentiation stage* in which neuronal progenitors appear as clusters of cells on top of an astrocytic cell monolayer.

3.2.1 Proliferation Stage

1. Plate the cells at a density of $2\text{–}3 \times 10^5$ cells/ml in complete NSC medium supplemented with 20 ng/ml EGF, 10 ng/ml b-FGF, and 5 % heat-inactivated FCS in an appropriate size tissue culture flask. Use 5 ml medium for T25, 20 ml for T75, and 40 ml for T175 flasks.
2. Incubate the NBA culture flasks in a 37 °C humidified incubator with 5 % CO_2 for 3–4 days. During this stage NSC progeny attach to the substrate and proliferate as a monolayer of cells (Fig. 1a). After 3–4 days of cell proliferation, the NBA culture becomes 90–95 % confluent (*see* **Note 4**).

3.2.2 Differentiation Stage

1. When the NBA culture becomes 90–95 % confluent, remove the medium from each NBA culture flask and replace it with freshly prepared complete NSC medium supplemented with 5 % FCS without any growth factors.

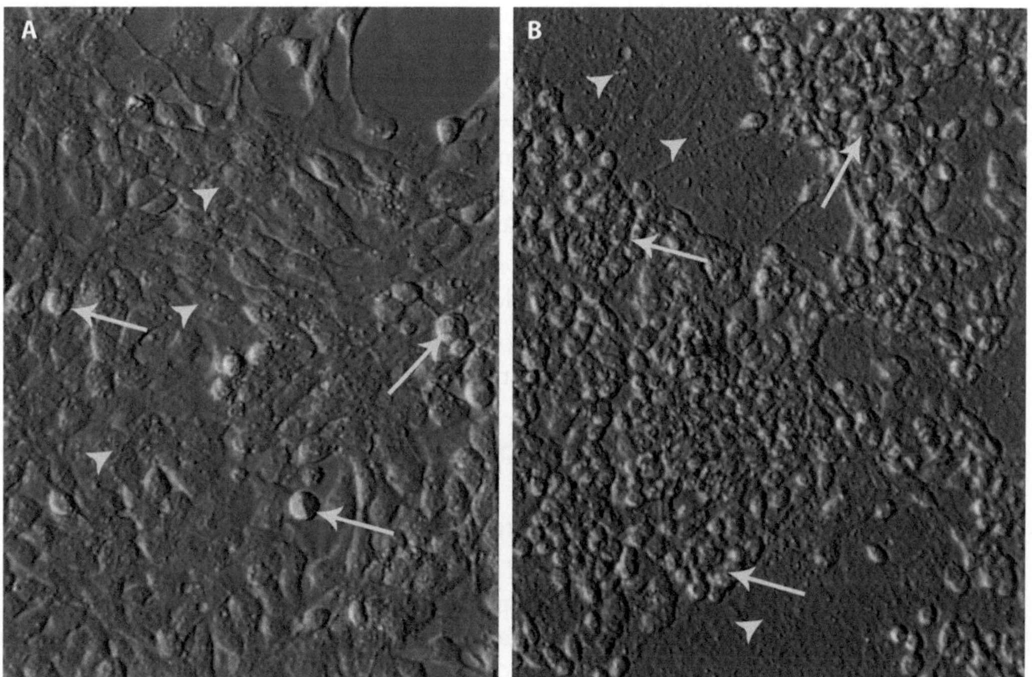

Fig. 1 Neuroblast assay differentiating culture of neural stem cells: Representative phase contrast pictures on day 4 of (**a**) proliferative and (**b**) differentiation stages. Note the flat astrocytic cells monolayer (*arrowheads*) underneath and the round neuronal progenitor cells (*arrows*) on *top*

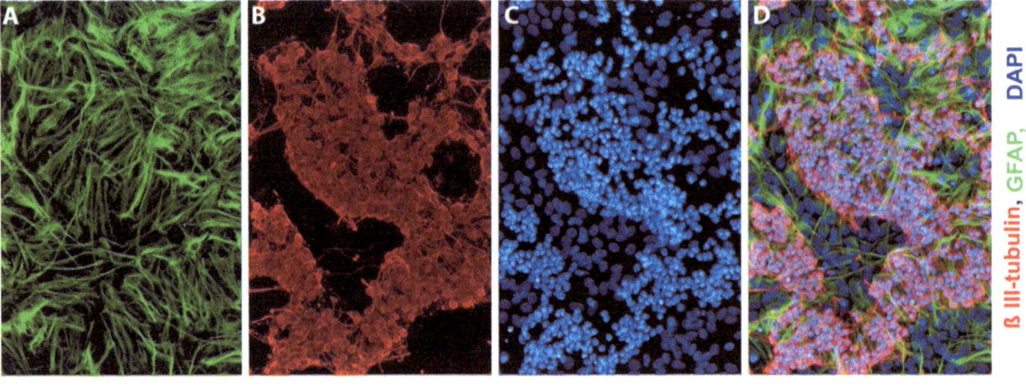

Fig. 2 Immunofluorescent staining of the neuroblast assay culture on day 4 of differentiation stage: (**a**) GFAP expressing astrocytes, (**b**). βIII tubulin positive clusters of neuronal progenitor cells, (**c**) DAPI counterstained nuclei, and (**d**) Merged picture. Note the size difference between astrocytes and neuronal cells

2. Incubate the cultures in a 37 °C humidified incubator with 5 % CO_2 for another 4–5 days. During this phase, neuronal progenitors appear on top of an astrocytic monolayer and proliferate to make colonies of immature neuroblast cells (Figs. 1b and 2).
3. Proceed to flow cytometry on day 4–5 of differentiation stage for isolation of the neuroblast cells (*see* **Note 5**).

3.3 Cell Preparation for Flow Cytometry

3.3.1 Preparing Single Cells from the NBA Culture

1. Wash the NBA culture with a sufficient amount of sterile PBS (2 ml for T25, 4 ml for T75, and 8 ml for T175 flask) to remove FCS from the culture (*see* **Note 6**).
2. Add a sufficient amount of pre-warmed 0.05 % Trypsin–EDTA (2 ml for T25, 4 ml for T75, and 8 ml for T175 flask) to each flask to cover the culture and incubate for 1–2 min in a 37 °C humidified incubator.
3. While holding the flask with one hand, strike firmly with the other hand to dislodge the cells from the bottom of the flask.
4. Add an equal amount of soybean trypsin inhibitor to the flask to stop trypsin activity (*see* **Note 7**).
5. Gently pipette the cell suspension up and down to ensure trypsin inactivation and to achieve a homogenous single cell suspension.
6. Pass the dissociated single cell suspension through a 40-µm size mesh filter to remove non-dissociated clumps.
7. Pellet the cell suspension at 800 rpm ($110 \times g$) for 5 min; discard the supernatant and resuspend the cells in an appropriate amount of complete NSC medium.
8. Proceed to isolate cells based on their physical properties as described in Subheading 3.4 or live cell immunostaining as appeared in Subheading 3.3.2.

3.3.2 Live Cell Immunostaining

Harvested single cells from the NBA culture can be stained for PSA-NCAM (a marker of early immature neuronal progenitors) to achieve neuronal progenitors with a higher purity.

1. Perform a cell count and divide the NBA single cell suspension into four groups: Cells alone, Cells plus PI, Isotype control, and Immunolabeling (*see* **Note 8**).
2. Pellet the cells to be stained at 800 rpm ($110\times g$) for 5 min, discard the supernatant and resuspend the cells in 100 μl of complete NSC medium.
3. Add Phycoerythrin (PE)-conjugated anti-PSA-NCAM antibody at a ratio of 10 μl/5–10×10^6 cells to the cells suspension to be stained for PSA-NCAM. In the same way, add PE-conjugated mouse IgG1 isotype control antibody to the isotype control cell suspension (*see* **Note 9**).
4. Mix the cell suspension gently and incubate in a dark place at room temperature for 10–15 min.
5. Pellet the cell suspensions (PSA-NCAM and Isotype control tubes) at 800 rpm ($110\times g$) for 5 min, discard the supernatant and resuspend the cells in 1 ml of complete NSC medium. Repeat this step 2–3 times to remove excess unbounded antibodies from the samples.
6. Resuspend the cells in an appropriate volume of NSC medium in 15 ml sterile falcon tubes (*see* **Note 10**) and add propidium iodide (PI, 500 μg/ml in PBS) at a concentration of 1 μl/ml of cell suspension only to the cells plus PI, isotype control, and PSA-NCAM-stained samples to exclude dead cells when analyzing by flow cytometry.
7. Prepare three 15 ml sterile falcon tubes with 1 ml of complete NSC medium supplemented with 1–2 % FCS to collect sorted cells from different cell populations.

3.4 Purification of Neural Cell Populations from Dissociated NBA Culture Using Fluorescent Activated Cell Sorting (FACS)

3.4.1 Setting Gates and Sorting

1. Run the cell suspension from the cells alone group through the FACS machine.
2. First, plot the cells (events) based on their size and internal complexity (forward scatter (FSC) light versus side scatter (SSC) light, respectively) to distinguish different cell populations (Fig. 3a). Adjust voltage parameters for FSC and SSC to see all acquired events in the flow cytometry plot.
3. Now plot the cells first based on side scatter area (SSC-A) versus side scatter pulse width (SSC-W) (Fig. 3b) and gate the resulting single cell population as population 1 (P1).
4. Then plot the P1 cells based on forward scatter area (FSC-A) versus forward scatter pulse width (FSC-W) (Fig. 3c) and gate the single cells as population 2 (P2). By excluding events with high FSC pulse width and high SSC pulse width, clumps, or doublets are excluded.

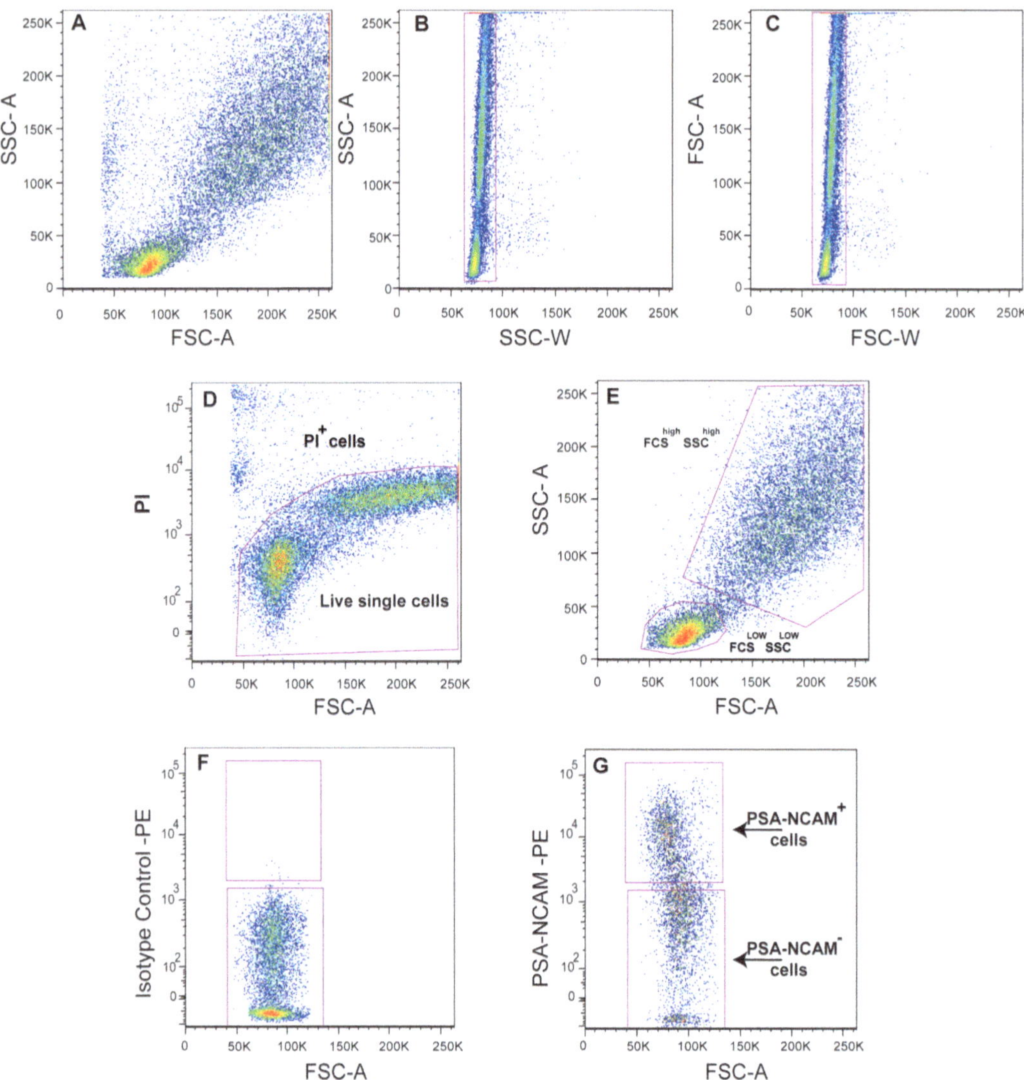

Fig. 3 Representative sort plots: (**a**) FSC vs. SSC sort plot before exclusion of doublets and dead cells, (**b**, **c**) two subsequent gates based on FSC and SSC pulse width to exclude doublets and clumps, (**d**) using propidium iodide immunoreactivity to remove dead and damaged cells, (**e**) FSC vs. SSC plot after removing doublets and dead cells, displaying the two main populations defined as the FSC^{low} SSC^{low} and the FSC^{high} SSC^{high} populations, (**f**) isotype control, and (**g**) PSA-NCAM-stained cells sort plots to select PSA immunoreactive cells from the FSC^{low} SSC^{low} population

5. To exclude dead or damaged cells, plot the single cells from P2 based on PI reactivity versus FSC-A and gate the single live cell population (Fig. 3d).
6. After exclusion of doublets, dead and damaged cells, again plot the single live cells based on FSC versus SSC and gate the two main cell populations as FSC^{low} SSC^{low} (P3) and FSC^{high} SSC^{high} (P4) cell populations (Fig. 3e). Proceed to cell sorting if you want to isolate cells only based on FSC versus SSC properties.

7. To sort PSA-NCAM immunoreactive neuronal progenitor cells, first plot the FSC^{low} SSC^{low} population from the control (cells plus PI and isotype control) groups based on PE immunoreactivity versus FSC-A and gate the negative cells as P5, and then run the cells from the PSA-NCAM stained group and gate the positive cells as population 6 (P6) (Fig. 3f, g).
8. Now sort the cells of interest into 15 ml sterile tissue culture tubes containing 1 ml NSC medium (*see* **Note 11**).

3.4.2 Plating the Sorted Cells

1. Pellet sorted cells at 1,200 rpm (240 × *g*) for 5 min.
2. Remove the supernatant and the resuspend the cells in an appropriate amount of medium (depending on the pellet size) and perform a cell count using Trypan blue exclusion.
3. Plate the cells in Poly-L-Ornithine-coated 96-well plates (*see* **Note 12**) at a density of 20–30 × 10^3 cells/well in 200–250 μl of complete NSC medium supplemented with 5 % FCS and 20 ng/ml human recombinant BMP-4.

3.4.3 Immunostaining of the Sorted Cells

1. Fix the cells at different time points after sort by adding 100 μl of cold 4 % PFA (in PBS, PH 7.2) to each well of 96-well plates and incubate for 15–20 min at room temperature.
2. Remove the PFA and gently rinse the sample 2–3 times using 100 μl/well of PBS each time.
3. Dispense 50–100 μl of the primary antibody solution to each well and incubate the samples at room temperature for 60–90 min (or overnight at 4 °C).
4. Gently remove the primary antibody and rinse the samples 2–3 times with PBS.
5. Incubate the samples with 50–100 μl/well of an appropriate fluorochrome-conjugated secondary antibody solution including DAPI (1:1,000, for nuclear counterstaining) for 45–60 min at room temperature in dark.
6. Gently remove the secondary antibody and rinse the samples 2–3 times with PBS.
7. Add 50–100 μl of PBS to each well and visualize the immunostained samples using a fluorescent microscope with appropriate filters (Figs. 4 and 5).

4 Notes

1. Do not let the neurospheres grow beyond 250 μm in diameter. Large and dark centered (overgrown) neurospheres will result in NSC progeny with decreased neurogenic ability upon differentiation.

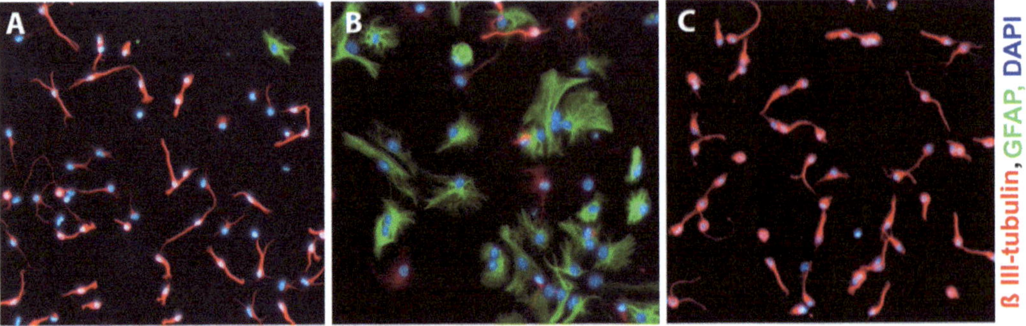

Fig. 4 Micrographs representing different cell population on day 1 after sort. (**a**) FSC^{low} SSC^{low} population of which the majority (greater than 75 %) are ß III tubulin positive neurons. (**b**) FSC^{high} SSC^{high} population of which the majority are GFAP expressing astrocytes. (**c**) PSA-NCAM^{+} sorted cells from FSC^{low} SSC^{low} population of which almost all of the sorted cells are neurons

2. Avoid leaving neurospheres in trypsin–EDTA for more than 3 min. Over-trypsinization increases cell death and reduces neuronal differentiation of the harvested NSC progeny.
3. Avoid pipetting the neurosphere suspension in trypsin–EDTA before adding trypsin inhibitor. This causes membrane rupture and cell lysis. Also avoid vigorous and numerous pipetting after addition of trypsin inhibitor as this might increase cell death and reduce neuronal differentiation of the harvested NSC progeny.
4. Do not let the proliferating NBA culture become overconfluent. Always switch the culture medium when it reaches around 90–95 % confluency.
5. On the night before flow cytometry, it is highly recommended to completely change the medium of the NBA culture with fresh complete NSC medium supplemented with 5 % FCS. This medium will become conditioned with many unidentified trophic and differentiation factors secreted from the astrocytes that can be used to plate isolated neuronal progenitor cells. The astrocyte-conditioned medium helps the survival and differentiation of the purified neurons.
6. FCS blocks Trypsin–EDTA and alternatively can be used to quench trypsin activity.
7. Alternatively, the supernatant culture medium from each flask, which contains 5 % FCS could be used to quench trypsin activity.
8. For Cells alone and Isotype control groups between $0.5–2 \times 10^6$ cells are needed but there is no limitation for Cells plus PI and Immunolabeling groups.
9. To achieve optimal staining follow the instructions on the antibody specification sheet regarding the concentration of the antibody to be used.

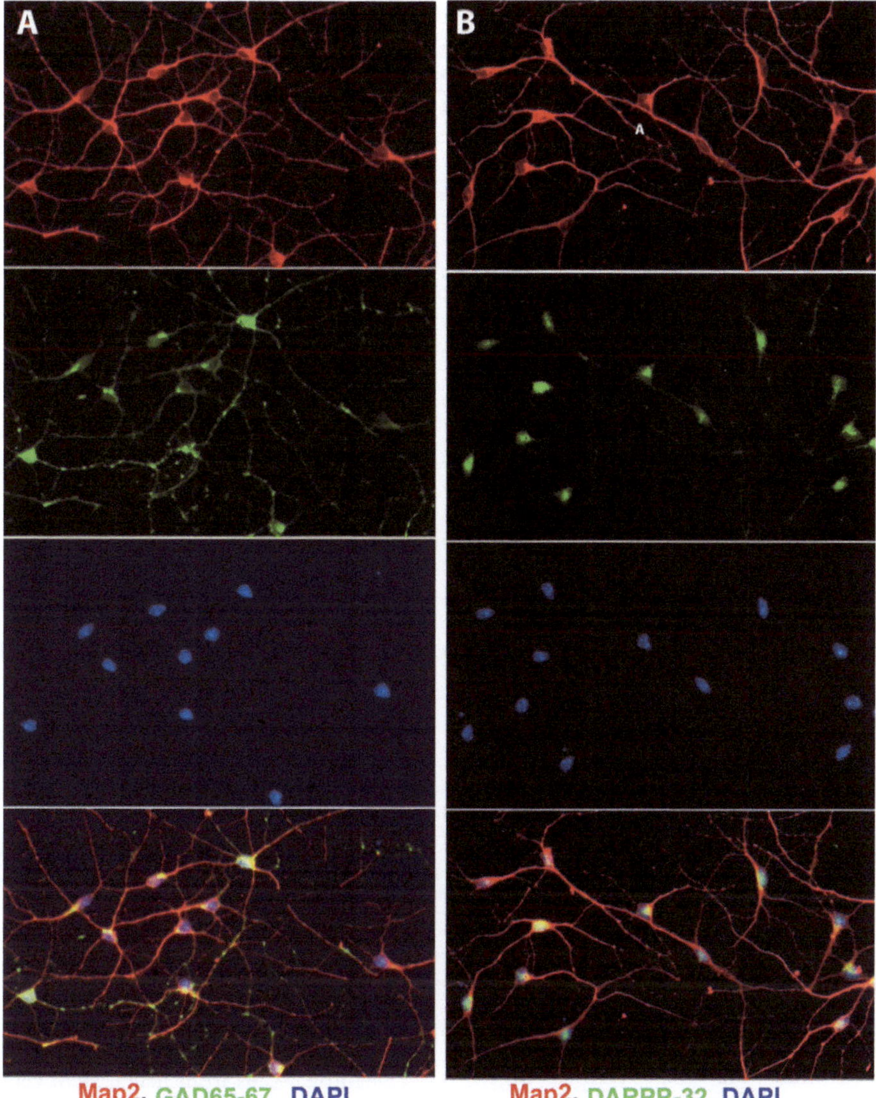

Fig. 5 Micrographs showing that sorted immature neurons differentiate into GABAergic (**a**) neurons and express DARPP-32 (**b**), a marker for medium spiny neurons

10. Cell density should not exceed $2–3 \times 10^6$ cells/ml. High cell density might cause stream blockage in the flow cytometry machine.
11. Use phosphate-buffered saline (PBS) or any other appropriate solution depending on manufacturer instructions, as the sheath fluid at 28 PSI through a 90 µm nozzle. Also set differential pressure on the system such that the sort trigger rate does not exceed 2,500 events/s.
12. Add 100 µl of Poly-L-Ornithine working solution (1.5 ml Poly-O and 8.5 ml of sterile PBS) to each well of 96-well plates

and incubate the plates in a 37 °C incubator for at least 1 h. Then, remove the Poly-O and wash each well three times with sterile PBS (let the PBS remain in the well for 10–15 min each time).

Acknowledgments

This work was supported by the Overstreet Foundation.

References

1. Reynolds BA, Weiss S (1992) Generation of neurons and astrocytes from isolated cells of the adult mammalian central nervous system. Science 255:1707
2. Azari H et al (2010) Isolation and expansion of the adult mouse neural stem cells using the neurosphere assay. J Vis Exp 45:e2393
3. Azari H et al (2011) Establishing embryonic mouse neural stem cell culture using the neurosphere assay. J Vis Exp 47:e2457
4. Siebzehnrubl FA et al (2011) Isolation and characterization of adult neural stem cells. Methods Mol Biol 750:61
5. Reynolds BA, Tetzlaff W, Weiss S (1992) A multipotent EGF-responsive striatal embryonic progenitor cell produces neurons and astrocytes. J Neurosci 12:4565
6. Suslov ON et al (2002) Neural stem cell heterogeneity demonstrated by molecular phenotyping of clonal neurospheres. Proc Natl Acad Sci U S A 99:14506
7. Bez A et al (2003) Neurosphere and neurosphere-forming cells: morphological and ultrastructural characterization. Brain Res 993:18
8. Taupin P (2005) Adult neurogenesis in the mammalian central nervous system: functionality and potential clinical interest. Med Sci Monit 11:RA247
9. Hofstetter CP et al (2005) Allodynia limits the usefulness of intraspinal neural stem cell grafts; directed differentiation improves outcome. Nat Neurosci 8:346
10. Raedt R et al (2009) Unconditioned adult-derived neurosphere cells mainly differentiate towards astrocytes upon transplantation in sclerotic rat hippocampus. Epilepsy Res 87:148
11. Amariglio N et al (2009) Donor-Derived Brain Tumor Following Neural Stem Cell Transplantation in an Ataxia Telangiectasia Patient. PLoS Med 6:e1000029
12. Rozental R et al (1995) Purification of cell-populations from human fetal brain using flow cytometric techniques. Brain Res Dev Brain Res 85:161
13. Schmandt T et al (2005) High-purity lineage selection of embryonic stem cell-derived neurons. Stem Cells Dev 14:55
14. Tsagias N et al (2011) Isolation of mesenchymal stem cells using the total length of umbilical cord for transplantation purposes. Transfus Med 21:253
15. Azari H et al (2011) Purification of immature neuronal cells from neural stem cell progeny. PLoS One 6:e20941
16. Azari H et al (2012) The neuroblast assay: an assay for the generation and enrichment of neuronal progenitor cells from differentiating neural stem cell progeny using flow cytometry. J Vis Exp 62:e3712

Chapter 10

Isolation of Adult Stem Cells from the Human Olfactory Mucosa

François Féron, Chris Perry, Stéphane D. Girard, and Alan Mackay-Sim

Abstract

The olfactory mucosa, located in the nasal cavity, is the only nervous tissue that is exposed to the external environment and easily accessible in every living individual. In addition, this organ is home of a continuing neurogenesis that is sustained by a large population of stem cells. Here, we describe a method for biopsy of olfactory mucosa from human nasal cavities and isolating multipotent adult stem cells that can be used to either identify biomarkers in brain disorders or repair the pathological/traumatized nervous system.

Key words Stem cell, Nose, Brain, Neuron, Cell therapy, Disease biomarkers, Neurosphere

1 Introduction

Molecular anomalies in brain diseases are usually studied using nervous tissue samples collected *postmortem*. However, this material has numerous limitations, including lengthened delays between death and autopsy. In contrast, the olfactory mucosa is readily accessible in every living individual and can be biopsied safely without any loss of sense of smell. Accordingly, the olfactory mucosa provides an "open window" in the adult human through which one can study developmental (e.g., autism, schizophrenia) [1–5] or neurodegenerative (e.g., Parkinson, Alzheimer) diseases [4, 6–8]. Patient-derived olfactory mucosa can be used for comparative molecular studies [4, 9], in vitro experiments on neurogenesis [3, 10], and cell biology [7, 11].

The olfactory epithelium is also a nervous tissue that produces new neurons every day to replace those that are damaged by environmental insults. This permanent neurogenesis is sustained by progenitors but also stem cells residing within both compartments of the human mucosa, namely the neuroepithelium and the underlying *lamina propria* [12–14]. We developed a method to isolate

Brent A. Reynolds and Loic P. Deleyrolle (eds.), *Neural Progenitor Cells: Methods and Protocols*, Methods in Molecular Biology, vol. 1059, DOI 10.1007/978-1-62703-574-3_10, © Springer Science+Business Media New York 2013

the human *lamina propria* from which we generated olfactory neurosphere derived (ONS) cell lines [4]. In a comparative study, we demonstrated that ONS are closely related to bone marrow mesenchymal stem cells (BM-MSC) and we named them olfactory ecto-mesenchymal stem cells (OE-MSC) [15].

We performed studies dedicated to unveil new candidate genes in schizophrenia, Parkinson's disease, and familial dysautonomia [4, 7, 9] and to understand gene regulation in disease states [16]. We and others also showed that human nasal olfactory stem cells are promising candidates for cell therapy, in animal models of paraplegia [17], cochlear damage [18], Parkinson's disease [19], intervertebral disc injury [20], or amnesia [21].

In this study, we present methods to biopsy olfactory mucosa in humans [22]. After collection, the *lamina propria* is enzymatically separated from the epithelium and stem cells are purified using an enzymatic or a nonenzymatic method. Purified nasal olfactory stem cells can then be either grown in large numbers and banked in liquid nitrogen or induced to form spheres or differentiated into neural cells. These stem cells can also be used for comparative molecular studies or cell therapy experiments.

2 Materials

2.1 Collection of Olfactory Mucosa

1. Rigid endoscope, Karl Storz™ or Richard Wolf Medical™.
2. Lidocaine/cocaine.
3. Epinephrine/phenylephrine/metazoline.
4. Throughput ethmoid forceps, Karl Storz™ or Richard Wolf Medical™.
5. Freer's knife.
6. Reabsorbable nasal dressing, Nasopore™.
7. DMEM/HAM F12 culture medium supplemented with fetal bovine serum (FBS) and antibiotics (gentamycin or a cocktail of penicillin/streptomycin).

2.2 Isolation of Olfactory Stem Cells

1. Dispase II (Roche), diluted at 2.4 IU/ml in DMEM/HAM F12 culture medium.
2. Dissecting microscope.
3. Micro spatula, FST.
4. Collagenase H (Sigma-Aldrich), diluted at 0.25 mg/ml in DMEM/HAM F12 culture medium.
5. Ca-free/Mg-free PBS.
6. DMEM/HAM F12 culture medium supplemented with FBS and 1 % streptomycin/penicillin.
7. Glass coverslip, Knittel Glaser.

2.3 Neurosphere Formation and Neuronal Differentiation

1. Poly-L-lysine.
2. Sphere-inducing culture medium: DMEM/HAM F12 supplemented with insulin, transferrin, selenium (ITS-X, 1 %), EGF (50 ng/ml), FGF2 (25–50 ng/ml), and antibiotics.
3. Neuron-differentiating culture medium: Neurobasal medium containing B-27 serum-free supplement, glutamine (2 mM), glutamate (0.025 mM), and antibiotics.

3 Methods

3.1 Collection of Olfactory Mucosa

1. This procedure should be carried out by an Ear Nose and Throat surgeon (otorhinolaryngologist), who is aware of the anatomy, to avoid complications such as cerebrospinal fluid leak and to handle the potential complication of a potential severe life-threatening nose bleed. This procedure should be done in accordance with and with the approval of the local research ethics committee and the patient should sign an informed consent.
2. The procedure can easily be done under a general anesthetic but it is quick and painless and safe to do under local anesthetic. The surgeon should be comfortable using a 0°or 30° rigid endoscope of either 2.7 or 4 mm diameter. The surgeon should inspect both nasal cavities and assess the presence or absence of nasal polyps or any inflammatory lesions which may be present. The presence of these may interfere with the growth of the olfactory cells. However much of the procedure can be performed by the surgeon using a headlight.
3. The surgeon inspects both sides of the nasal cavity. It is usually best to spray both sides of the nasal cavity with a vasoconstricting solution such as Metazoline or Phenylephrine. These sprays take 2–3 min to shrink the nasal mucosa to allow better inspection of the upper nose. The best position to take a biopsy is on the nasal septum approximately half way back along the length of the middle turbinate, using the middle turbinate as a landmark for the otherwise featureless nasal septum. Biopsies are best taken from the septum about a centimeter from the roof of the nasal cavity, the cribriform plate. The better side is usually the one most easily seen into. It is also usually the side the septum is bent away from. There is often a deviation of the septum to one side more than the other in the superior part of the nose.
4. The nasal cavity is prepared with a local anesthetic solution which could be Lidocaine with adrenalin (epinephrin) topical solution or 10 % cocaine solution. It takes a number of applications of the soaked cotton tipped stick painted over the nasal

cavity over a 10-min period. The surgeon then uses a long, fine bore needle to raise a small bleb of Lidocaine with adrenalin local anesthetic into the biopsy site. This bleb should be a couple of millimeter thick spread over an area of about a centimeter.

5. The surgeon then uses a number 15 scalpel blade to make a small vertical cut 6–8 mm in the nasal septum in the front of the area raised as a bleb by the local anesthetic injection. A semi-sharp dissector such as a Freer's knife raises a small sub-periosteal flap back for 2–3 mm. A small through-cut ethmoid forcep or similar small ENT grabbing forcep is put into the nose and a small piece of the raised mucosa with underlying periosteum is snipped off. This piece of mucosa should be at least 2 mm × 1 mm.
6. To prevent later significant bleeding from this small area of trauma in the upper nasal septum, place over the biopsy site a small pack of Nasopore™ (a dissolving nasal pack). This pack is wedged in the upper nose between the middle turbinate and the septum over the biopsy site. The pack is covered in an antibiotic ointment to minimize the chance of toxic shock. It acts as a framework for blood clot and stops the exposed biopsied area drying, cracking, and bleeding. Without coverage of this small area of trauma in the upper part of the nose, as the vasoconstrictor agents wear off, severe and potentially life-threatening hemorrhage could occur. This small piece of dissolving pack (sized about 2 cm × 1 cm × 1 cm) is normally not noticed by the patient. It dissolves over about a week without them being aware it is there and effectively prevents reactionary and secondary hemorrhage.
7. The olfactory biopsy is then transferred, using a sterile needle, into a sterile 2 ml tube filled with 1 ml of cold Dulbecco's Modified Eagle's Medium/Ham F-12 (DMEM/F12) containing 10 % FBS and 1 % streptomycin/penicillin. This culture medium will be used through the whole procedure, except when generating spheres or during the differentiation processes. Tip the tube upside down to make sure that the biopsy is immersed in the culture medium.
8. Insert the tube in a refrigerated container and transport it to the research laboratory. At this stage, the biopsy can be used per se for comparative molecular studies focused on specific brain diseases or processed for generating stem cells (*see* **Note 1**).

3.2 Isolation of the Olfactory Epithelium from the Lamina Propria

1. Wash the biopsies in DMEM/HAM F12. Incubate the biopsies in a Petri dish filled with 1 ml of Dispase II solution, for 1 h at 37 °C.
2. Next, under a dissecting microscope with a diffracted inverted light, the olfactory epithelium is removed from the underlying lamina propria using a micro spatula (*see* **Note 2**).

3. The olfactory epithelium is thinner and looks translucent over a black background compared to the lamina propria which is striped orange/brown. Over a white background, the epithelium looks gray and the lamina propria, brown.
4. Once separated, transfer the lamina propria and olfactory epithelium into separate Petri dish filled with DMEM/HAM F12.

3.3 Culture of Olfactory Stem Cells from the Lamina Propria

1. Slice the lamina propria into 3–4 pieces with a thickness ranging from 200 to 500 μm.
2. Insert each strip into its own 2 cm diameter culture dish and cover the tissue with sterile 1.3 cm diameter glass cover slips.
3. Add 500 μl of DMEM/F12 culture medium to each culture dish.
4. Renew the culture medium every 2–3 days.
5. 5–7 days after, stem cells will begin to invade the culture dish and after 2 weeks they should be confluent. When confluency is reached, passage and transfer the cells to culture flasks (*see* **Note 3**).

3.4 Neurosphere Formation

1. The biopsies are received in cold DMEM/F12.
2. Olfactory epithelium and lamina propria are separated (above).
3. Olfactory epithelial suspensions are created by mechanical trituration through a Pasteur pipette.
4. Lamina propria cell suspensions were generated by digestion with Collagenase H (10 min, 37 °C) accompanied by mechanical trituration as described [13].
5. Dissociated epithelial and lamina propria cells are recombined and grown for 3 days in DMEM/F12, according to the next three steps.
6. Prepare flasks by incubating them for 2 h at 37 °C with poly-L-lysine (1–5 μg/cm^2) (*see* **Note 4**).
7. Passage the olfactory cells using trypsin or similar and plate the cells at a density of 16,000 cells/cm^2 in the prepared flasks.
8. At Day 1 post-plating, collect the supernatant in which undissociated bits of lamina propria float, centrifuge it at 200 × *g* and mechanically dissociate the floating lamina using a Pasteur pipette before replating the cells in a new well. The first well containing the already attached cells is filled with fresh culture medium.
9. Every 2 days, feed the cells with 0.2 ml/cm^2 of the sphere-inducing culture medium.
10. Neurospheres form initially from cell clusters attached to the culture dish surface but detach when they reached about 100 mm in diameter. Free-floating neurospheres are harvested every second day from the medium change.

11. Neurospheres are then used for differentiation or replated for stem cell banking as "neurosphere-derived cells" (ONS cells; ref. 4) (*see* **Note 4**).

3.5 Neurosphere-Derived Cells and Cell Banking

1. Neurospheres are dissociated with trypsin, replated at 4,000 cells/cm^2 into 75 cm^2 flasks, and cultured in DMEM/F12 culture medium. These ONS cells are then expanded by passage and banked down in aliquots after harvest by storage in liquid nitrogen with 90 % FBS and 10 % dimethyl sulfoxide.
2. For banking purposes, multiple aliquots are generated at early passage, nominated "master stocks". Master stock aliquots are expanded and aliquoted as "distribution stocks". Distribution stock aliquots are expanded and aliquoted as "experimental stocks".
3. Frozen aliquots of experimental stocks are used as the starting point for all experiments allowing multiple experiments to be performed on cells of similar passage number, normally restricted to passage 6–10.
4. Frozen aliquots are thawed and grown under standard conditions on tissue culture plastic in DMEM/F12 culture medium at 37 °C and 5 % CO_2.

3.6 Neuronal Differentiation of Neurospheres

1. Neurospheres are dissociated with trypsin and cultivated for 21 days in neuron-differentiating culture medium.
2. Make medium changes every 3 days. Neuron-like cells should appear after 2–3 weeks (*see* **Note 5**).

4 Notes

1. The protocol for biopsy is the same used for generating olfactory ensheathing cell cultures [23]. Larger biopsies are taken under general anesthesia for clinical application [24].
2. The protocol is slightly modified in order to generate olfactory neurons in vitro [3, 10, 22]. For that purpose, the neuro-epithelium is not removed and the whole olfactory mucosa is sliced with a McIlwain chopper (200 μm thickness). Each explant is plated in a dish, partially dried for 1 h and then rehydrated with DMEM/HAMF12 culture medium. During the first days post-plating, epithelial and mesenchymal cells grow out of the explant. Then, neuron progenitors will migrate on the top of this cell layer and differentiate into neurons.
3. Purification of olfactory stem cells can be achieved using flow cytometry. Specific surface markers can be retrieved from the list published in the characterization paper [15].

4. Neurospheres can also be generated, any time, from adherent olfactory stem cells cultivated on plastic and in serum-containing medium. Flasks are pretreated with Poly-L-Lysine. Stem cells are passaged and cultivated in sphere-inducing culture medium.
5. As a member of the mesenchymal stem cell superfamily, OE-MSCs are able, under appropriate culture conditions, to differentiate into adipocytes, chondrocytes, osteocytes, and myocytes [15, 20].

Acknowledgments

This work was financially supported by ANR (Agence nationale de la Recherche), AFM (Association Française contre les Myopathies), FEDER in PACA and IRME (Institut de Recherche sur la Moelle épinière et l'Encéphale), the Australian Department of Health and Ageing, and the National Health and Medical Research Council of Australia.

References

1. Abrams MT et al (1999) FMRl gene expression in olfactory neuroblasts from two males with fragile X syndrome. Am J Med Genet 82:25–30
2. Arnold SE et al (2001) Dysregulation of olfactory receptor neuron lineage in schizophrenia. Arch Gen Psychiatry 58:829–835
3. Feron F et al (1999) Altered adhesion, proliferation and death in neural cultures from adults with schizophrenia. Schizophr Res 40:211–218
4. Matigian N et al (2010) Disease-specific, neurosphere-derived cells as models for brain disorders. Dis Model Mech 3:785–798
5. Ronnett GV et al (2003) Olfactory biopsies demonstrate a defect in neuronal development in Rett's syndrome. Ann Neurol 54:206–218
6. Arnold SE et al (2010) Olfactory epithelium amyloid-beta and paired helical filament-tau pathology in Alzheimer disease. Ann Neurol 67:462–469
7. Cook AL et al (2011) NRF2 activation restores disease related metabolic deficiencies in olfactory neurosphere-derived cells from patients with sporadic Parkinson's disease. PLoS One 6:e21907
8. Wolozin B et al (1993) A.E. Bennett Research Award, 1993. Olfactory neuroblasts from Alzheimer donors: studies on APP processing and cell regulation. Biol Psychiatry 34: 824–838
9. Boone N et al (2010) Olfactory stem cells, a new cellular model for studying molecular mechanisms underlying familial dysautonomia. PLoS One 5:e15590
10. McCurdy RD et al (2006) Cell cycle alterations in biopsied olfactory neuroepithelium in schizophrenia and bipolar I disorder using cell culture and gene expression analyses. Schizophr Res 82:163–173
11. Ghanbari HA et al (2004) Oxidative damage in cultured human olfactory neurons from Alzheimer's disease patients. Aging Cell 3:41–44
12. Murrell W et al (1996) Neurogenesis in adult human. Neuroreport 7:1189–1194
13. Murrell W et al (2005) Multipotent stem cells from adult olfactory mucosa. Dev Dyn 233: 496–515
14. Roisen FJ et al (2001) Adult human olfactory stem cells. Brain Res 890:11–22
15. Delorme B et al (2010) The human nose harbors a niche of olfactory ectomesenchymal stem cells displaying neurogenic and osteogenic properties. Stem Cells Dev 19:853–866
16. Mar JC et al (2011) Variance of gene expression identifies altered network constraints in neurological disease. PLoS Genet 7:e1002207
17. Xiao M et al (2005) Human adult olfactory neural progenitors rescue axotomized rodent rubrospinal neurons and promote functional recovery. Exp Neurol 194:12–30

18. Pandit SR et al (2011) Functional effects of adult human olfactory stem cells on early-onset sensorineural hearing loss. Stem Cells 29:670–677
19. Murrell W et al (2008) Olfactory mucosa is a potential source for autologous stem cell therapy for Parkinson's disease. Stem Cells 26: 2183–2192
20. Murrell W et al (2009) Olfactory stem cells can be induced to express chondrogenic phenotype in a rat intervertebral disc injury model. Spine J 9:585–594
21. Nivet E et al (2011) Engraftment of human nasal olfactory stem cells restores neuroplasticity in mice with hippocampal lesions. J Clin Invest 121:2808–2820
22. Feron F et al (1998) New techniques for biopsy and culture of human olfactory epithelial neurons. Arch Otolaryngol Head Neck Surg 124:861–866
23. Bianco JI et al (2004) Neurotrophin 3 promotes purification and proliferation of olfactory ensheathing cells from human nose. Glia 45:111–123
24. Feron F et al (2005) Autologous olfactory ensheathing cell transplantation in human spinal cord injury. Brain 128:2951–2960

Part II

Stem Cells vs Progenitor Cells

Chapter 11

Enumerating Stem Cell Frequency: Neural Colony Forming Cell Assay

Sharon A. Louis and Carmen K.H. Mak

Abstract

Recent reports have highlighted several parameters of the neurosphere culture or assay system which render it unreliable as a quantitative in vitro assay for measuring neural stem cell (NSC) frequency. The single-step semi-solid based assay, the Neural Colony Forming Cell (NCFC) assay is an assay which was developed to overcome some of the limitations of the neurospheres assay in terms of accurately measuring NSC numbers. The NCFC assay allows the discrimination between NSCs and progenitors by the size of colonies they produce (i.e. their proliferative potential). The NCFC assay and other improved tissue culture tools offer further advances in the promising application of NSCs for therapeutic use.

Key words Murine, Embryonic neural stem cells, Subventricular zone cells, Neurospheres, Frequency neural stem cell, Culture, Stem cells

1 Introduction

The functional characterization of putative NSC often relies on in vitro culture systems which provide a retrospective readout of their function and frequencies. The neurosphere culture system [1] for example, can be used to demonstrate the cardinal stem cell properties of: (1) self-renewal over an extended period of time, (2) generation of a large number of progeny and (3) multi-lineage differentiation potential and remains the most frequently adopted method to enrich, expand and differentiate NSC [1–3].

Since its original discovery, many modifications to the neurosphere culture media and protocols have been introduced into the neural stem cell field by individual labs. The modifications have not been standardized across the field, and has lead to variations in protocols and media being used by different labs and thus contributing to the discrepancies in the results obtained within labs and between labs.

Brent A. Reynolds and Loic P. Deleyrolle (eds.), *Neural Progenitor Cells: Methods and Protocols*, Methods in Molecular Biology, vol. 1059, DOI 10.1007/978-1-62703-574-3_11,

The application of these non-standardized methodologies, have also been applied to quantification of NSC frequencies. Reynolds and Rietze [4] challenged one of the central tenets of the neurosphere assay (that all neurospheres are derived from a NSC) and concluded that an exclusively one-to-one relationship between neurosphere formation and NSCs does not exist, thereby, suggesting that this tenet is incorrect [4]. In addition several recent publications have also highlighted some of the limitations of the neurosphere culture system as a read-out for NSC numbers [5, 6].

To address some of the shortcomings of the neurosphere assay, an assay called the Neural Colony Forming Cell (NCFC) assay which is able to more accurately discriminate between neural stem and progenitor cells compared to the neurosphere assay was developed [7]. In the NCFC assay colony size is an indicator of proliferative potential and cells which form colonies >2 mm in diameter meet all the functional criteria for a NSC which includes the ability to self-renew, generate large numbers of progeny and maintain multipotency over an extended period of time. Cells which form colonies <2 mm lack self-renewal ability and are likely progenitor. The combination of low cell plating densities and use of the collagen semi-solid media in the NCFC assay ensures clonality by preventing cell motility, cell–cell or colony contacts and cell aggregation. The NCFC assay therefore provides a quantitative read-out for NSCs that is more accurate than measurements of neurosphere number or secondary/tertiary neurosphere formation ability that is routinely practiced in the field. The NCFC assay is therefore a useful tool to measure NSC and progenitor cell frequency not only in routine neural cultures but also in cell samples isolated directly from normal or degenerative CNS tissues and from tissues previously treated with exogenous compounds.

This chapter attempts to describe the useful features of the NCFC assay, the NCFC assay methodology and its application in the study of neural stem biology. The chapter also includes methods for obtaining cells from neurospheres and adherent monolayer cultures derived from mouse CNS cells which can be used as a start sample for performing the NCFC assay. With these complete set of accurate protocols and validated reagents, we attempt to facilitate standardization of neural stem cell and progenitor cell enumeration.

2 Materials

2.1 General Equipment

1. Biological safety cabinet (e.g., Canadian Cabinets) certified for Level II.
2. Low-speed centrifuge (e.g., Beckman TJ-6) equipped with biohazard containers.

3. 37 °C Incubator with humidity and gas control to maintain >95 % humidity and an atmosphere of 5 % CO_2 in air (e.g., Forma 3326).
4. Pipette-aid (e.g., Drummond Scientific).
5. Hemacytometer (e.g., Brightline).
6. Trypan blue (e.g., cat. no. 07050, STEMCELL Technologies Inc.).
7. Light microscope with 5× and 10× objectives for hemacytometer cell counts.
8. Inverted microscope with flatfield objectives and eyepieces to give object magnification of approx. ×20–×30, ×80, and ×125 (e.g., Nikon Diaphot TMD).
9. Pasteur glass pipettes, sterile.

2.2 Tissue Culture Equipment (Neurosphere and Adherent Cultures)

1. T-25 cm^2 tissue culture flask (Nunc or VWR) or T-162 cm^2 Flask (Corning).
2. Tubes, 17 mm × 100 mm polystyrene test tubes, sterile (e.g, BD Falcon™).
3. Tubes, 50 mL, polypropylene, sterile (e.g, BD Falcon™).
4. 24-Well culture dishes (e.g., Corning).
5. 6-Well culture dishes (e.g., Corning).

2.3 Media, Supplements and Associated Reagents for Neurosphere Cultures

2.3.1 Proliferation

The performance of media prepared in the laboratory is highly dependent on the quality and purity of the water and raw materials. If media is prepared in the laboratory, use only tissue culture grade materials and if necessary source various suppliers to determine the best quality reagents as there is significant batch to batch variability in some critical reagents. To avoid variability in media performance, STEMCELL provides optimized and standardized kits for the proliferation of neural cells (*see* **Note 1**).

1. NeuroCult® NSC Basal Media (Mouse) (STEMCELL Technologies Inc.).
2. NeuroCult® NSC Proliferation Supplements (STEMCELL Technologies Inc.).
3. *NeuroCult™ Basal-Proliferation Mix Media.* This media is prepared as follows: Thaw an aliquot of the NeuroCult® NSC Proliferation Supplements from **item 2.3**. Add 50 mL of the NeuroCult® NSC Proliferation Supplements to 450 mL of NeuroCult® NSC Basal Media (Mouse) from **item 1** to give a 1:10 dilution. The proliferation-supplemented neural culture media should be stored at 4 °C and used within 1 week.
4. Human recombinant epidermal growth factor (rhEGF) (STEMCELL Technologies Inc.). A stock solution of 10 μg/mL

of rhEGF is made up in 1 mL sterile diluent containing 10 mM acetic acid and at least 0.1 % bovine serum albumin, then adding 19 mL of the Basal-Proliferation Mix Media from **item 3** and stored as 1-mL aliquots at −20 °C until required for use.

5. Human recombinant fibroblast growth factor (rhFGF) (for adult mouse CNS cells; STEMCELL Technologies Inc.). A stock solution of 10 μg/mL of rhFGF is made up in PBS containing at least 0.1 % bovine serum albumin and stored as 1-mL aliquots at −20 °C until required for use.
6. 0.2 % Heparin (for adult mouse CNS cells). Mix 100 mg of heparin (Sigma-Aldrich) in 50 mL of distilled water. Filter sterilize. Store aliquots of 1 mL at 4 °C.
7. "*Complete*" *NSC Proliferation medium*. Add 2 μL of rhEGF to every 1 mL of the Basal-Proliferation Mix Media from **item 3** to give a final concentration of 20 ng/mL of EGF.
8. "*Complete*" *NSC Proliferation medium* (*Adult*). Add 2 μL of rhEGF, 1 μL of rhFGF and 1 μL of heparin to every 1 mL of the Basal-Proliferation Mix Media from **item 3** to give a final concentration of 20 ng/mL of EGF, 10 ng/mL bFGF and 0.0002 % heparin.
9. Accutase™ (cat. no. 07920. STEMCELL Technologies Inc.). Accutase™ is used to detach adherent monolayer cells from the flasks to harvest the cells. Following steps in Subheading 3.1.2.

2.4 Media, Supplements and Tissue Culture Equipment for Neural Colony Forming Cell Assay

Note that while we originally developed research grade media formulations for the NCFC assay, the reagents to perform this assay are now available through a commercial supplier. The Neural Colony Forming Assay called NeuroCult® NCFC Assay Kit (Mouse) is currently supplied as a ready-to-use kit from STEMCELL Technologies Inc.

1. NeuroCult® NCFC Serum-Free Medium without Cytokines.
2. NeuroCult® NSC Proliferation Supplements (Mouse).
3. NeuroCult® NSC Basal Medium (Mouse).
4. Collagen Solution (STEMCELL).
5. 35 mm Culture Dishes seven packs (ten dishes/pack) (STEMCELL).
6. Gridded Scoring Dishes (STEMCELL).
7. 40 μm cell strainer (STEMCELL Technologies Inc.).
8. 245 mm square bioassay dish (STEMCELL Technologies Inc.).

3 Methods

3.1 Cells from Primary Embryonic or Adult CNS Tissues and Neurosphere or Adherent Monolayer Cultures for Use in the NCFC Assay

3.1.1 Preparation of Cells from Primary Embryonic or Adult CNS Tissues

1. Resuspend dissected embryonic tissues in 1 mL of Basal-Proliferation Mix Media (Subheading 2.3, **item 3**).
2. Adult CNS cells are prepared according to standard lab protocols or using the NeuroCult™ Enzymatic Dissociation kit for Adult CNS tissues (*see* **Note 2**).
3. Using a Pasteur glass pipette or a plastic disposable tip attached to a P1000 micropipetor set at 1 mL; triturate the tissue for approximately three times (*see* **Note 3**).
4. Resuspend cells in a total volume of 10 mL Basal-Proliferation Mix Media.
5. Centrifuge the cells at 800 rpm ($110 \times g$) for 5 min. Remove supernatant and resuspend the cells with a brief trituration in 2 mL medium (*see* **Note 4**).
6. If undissociated tissue remains, allow the suspension to settle for 1–2 min and then pipette off the supernatant containing single cells into a fresh tube (*see* **Note 4**).
7. Filter single cell suspension over a 40 μm cell strainer.
8. Measure the precise volume and count cell numbers using a dilution in trypan blue (1/5 or 1/10 dilution) and hemacytometer.
9. *To set up NCFC assay Section 3.2 step 3a for Primary embryonic cells*: Dilute primary embryonic cells to 6.5×10^5 cells/mL in Complete NSC Proliferation Medium which will give a final cell plating density of 7,500 cells per 35 mm culture dish in a 25 μL volume. Proceed to **step 1** of Subheading 4.1 to set the NCFC assay.

3.1.2 Preparation of Primary Neurosphere Cultures from Embryonic CNS Cells

1. To set up primary neurospheres cultures of embryonic CNS cells from **step** 7 of Subheading 3.1, seed cells at a density of 2×10^6 cells per 10 mL or 80, 000 cells/cm^2 (T-25 cm^2 flask) or 8×10^6 cells in 40 mL media (T-162 cm^2 flask), in Complete NSC Proliferation Medium.
2. To set up neurospheres cultures of adult SVZ cells from Subheading 3.1.3, seed primary adult cells at 2×10^4 cells/cm^2 (1.9×10^5 total cells) in 6-well tissue culture dishes (Corning) or 2×10^4 cells/cm^2 (5×10^5 total cells) in a T-25 cm^2 flask (Nunc) containing Complete NeuroCult® NSC Proliferation Medium (Adult).

3.1.3 Preparation of Primary Adherent Monolayer Cultures from Embryonic CNS Cells

There are several publications indicating that neural stem cells and neural progenitors can also be maintained in adherent monolayer cultures [8, 9]. The NCFC assay can be used to enumerate the numbers of neural stem cells and neural progenitor cells in the adherent monolayer cultures (Fig. 1).

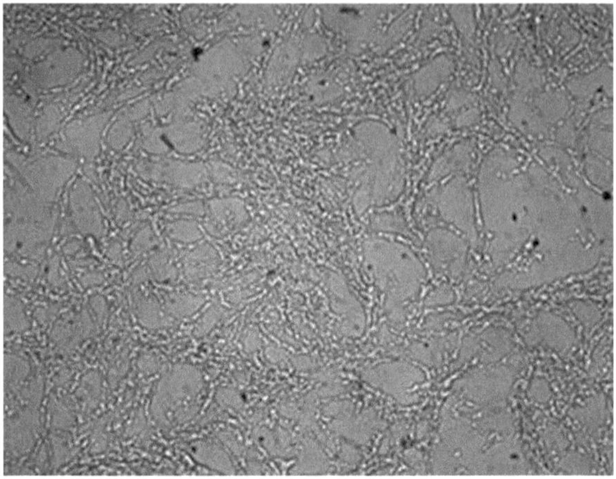

Fig. 1 Adherent cultures derived from adult SVZ cells. Cells cultured for 10 days in passage 2 have attached to the bottom of the vessel and formed a monolayer of cells which is about 70–80 % confluent

1. To set up primary adherent monolayer cultures of embryonic CNS cells, seed cells from **step 8** of Subheading 3.1.1 at a density of 7.6×10^5 cells per 3 mL (8×10^4 cells/cm^2) in a coated 6-well tissue culture dishes (*see* **Note 5**) (Corning) or 2×10^6 cells per 10 mL (8×10^4 cells/cm^2) in a coated T-25 cm^2 flask (*see* **Note 5**) (Nunc) containing "Complete" NSC Proliferation Medium.
2. To set up primary adherent monolayer cultures of adult SVZ cells, seed cells at 1.9×10^5 cells per 3 mL (2×10^4 cells/cm^2) in a coated 6-well tissue culture dishes (*see* **Note 5**) (Corning) or 5×10^5 cells (2×10^4 cells/cm^2) in a coated T-25 cm^2 flask (Nunc) containing "Complete" NSC Proliferation Medium (Adult).

3.1.4 Preparing Cells from Neurospheres Derived from Embryonic or Adult Mouse CNS Cells

1. Observe the neurosphere cultures under a microscope to determine if the neurospheres are ready for passaging (4–7 days) (Fig. 2). If neurospheres are attached to the culture flask, tapping the culture flask against the bench top should detach them.
2. Using a pre-wetted disposable pipette, remove medium with suspended neurosphere from a T-25 cm^2 flask and place in a 14 mL sterile tissue culture tube. If some cells remain attached to the flask, detach them by shooting a stream of media across the attached cells. Spin at 400 rpm ($75 \times g$) for 5 min.
3. Remove all the supernatant and resuspend cells in a maximum of 200 μL of complete NSC medium. The neurospheres in the pellet will need to be dissociated into a single cell suspension (*see* **Note 6**).
4. With a plastic disposable pipette tip attached to a P200 pipettor set at ~180 μL, triturate the neurospheres until single cell suspension is achieved (*see* **Note 7**).

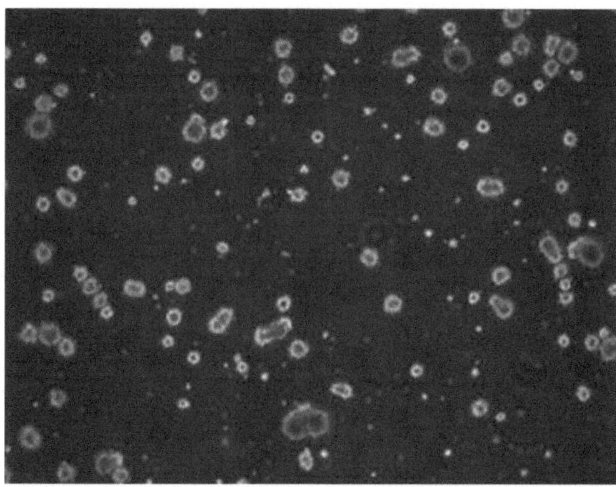

Fig. 2 Neurosphere cultures derived from adult SVZ cells. At 5–10 days in culture, SVZ cells formed healthy neurospheres which continue to grow in size and are phase contrast bright

5. Filter the single cell suspension through a 40 μm cell strainer to remove undissociated cell or clumps of cells.
6. Measure the precise volume and count cell numbers using a dilution in trypan blue (1/5 or 1/10 dilution) and hemacytometer.
7. To set up NCFC assay in section 3.2 step 3b *cells derived from neurosphere cultures*: Dilute these cells to 2.2×10^5 cells/mL which will give a final cell plating density of 2,500 cells per 35 mm culture dish in a 25 μL volume. Proceed to **step 1** of Subheading 3.2.

3.1.5 Obtaining Cells from Adherent Monolayer Derived from Embryonic or Adult Mouse CNS Cells

1. Observe the adherent monolayer cultures under a microscope to determine if the cultures are ready for passaging (4–7 days) (Fig. 1). The cultures should be passaged when they reach 60–80 % confluence.
2. Using a disposable pipette, remove medium from the culture vessel and discard. Add 2 mL PBS to each well of a 6-well plate or 5 mL PBS to each T-25 cm^2 flask. Swirl the culture vessel gently, remove the PBS and discard.
3. To harvest the cells, add 0.5 mL ACCUTASE™ to each well of a 6-well plate or 1 mL ACCUTASE™ to each T-25^2 flask. Incubate for 5 min at 37 °C.
4. Observe the culture to determine if the cells are starting to detach and detachment is complete. Tapping the culture vessels against the bench top should detach them.
5. Add 2 mL of "Complete" NSC Proliferation medium to each well of a 6-well plate or 5 mL of "Complete" NSC Proliferation medium to each T-25 cm^2 flask. Using the same pipette, collect all detached cells into a 14 mL tube.

6. Spin at 800 rpm ($110 \times g$) for 5 min.
7. Remove all supernatant and resuspend cells in a maximum of 200 μL of complete NSC medium.
8. With a plastic disposable pipette tip attached to a P200 pipettor set at ~180 μL, pipetting until single cell suspension is achieved (*see* **Note 7**).
9. Filter the single cell suspension through a 40 μm cell strainer to remove undissociated cell or clumps of cells.
10. Measure the precise volume and count cell numbers using a dilution in trypan blue (1/5 or 1/10 dilution) and hemacytometer.
11. To set up NCFC assay, dilute cells to 2.2×10^5 cells/mL which will give a final cell plating density of 2,500 cells per 35 mm culture dish in a 25 μL volume. Proceed to **step 4** in Subheading 3.2.

3.2 Enumeration of Neural Stem Cells and Neural Progenitor Cells Using the Neural Colony Forming Cell Assay (NCFCA)

The NeuroCult® NCFC Assay allows for the identification and discrimination of NSC and progenitor cells from mouse or rat primary CNS tissue and the cultured neurospheres or adherent monolayer cultures from these species based on their proliferative potentials (Fig. 3). Primary or cultured neural cells are suspended in serum-free medium containing optimized levels of growth supplements and recombinant cytokines. Collagen is then mixed with the cell-medium suspension and dispensed into 35 mm culture dishes. At the end of the 21 day culture period, clonally derived colonies of different sizes are scored.

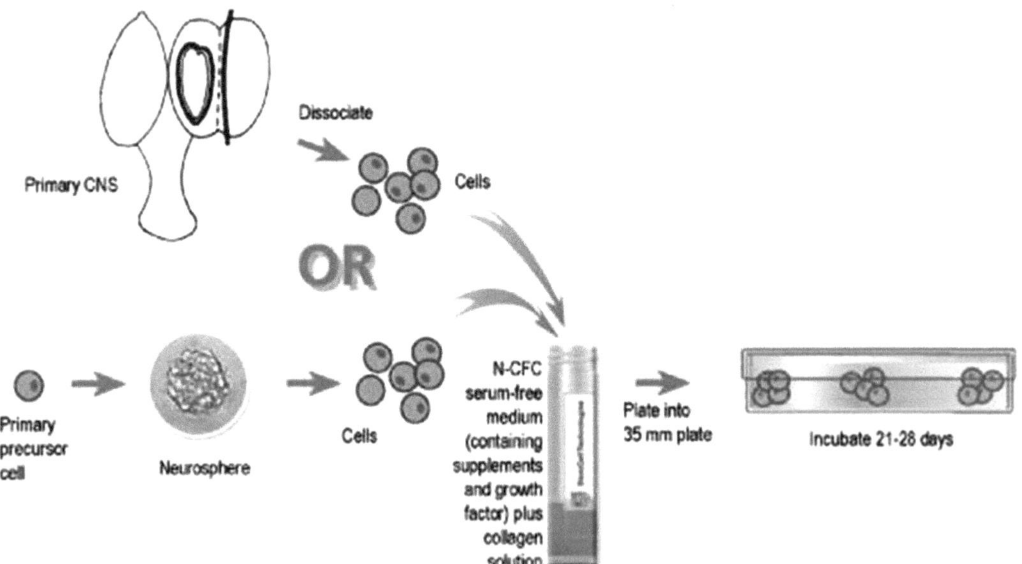

Fig. 3 The procedure to set up NCFC cultures from primary CNS tissue and cells derived from neurosphere cultures

1. Thaw bottles or aliquots of NeuroCult® NCFC Serum-Free Medium without Cytokines and NeuroCult® NSC Proliferation Supplements (Mouse) at room temperature or overnight at 4 °C.
2. Place thawed medium and supplements at 37 °C and the Collagen Solution on ice (*see* **Note 8**).
3. Filter the single cell suspension of neural cells (derived from cultured neurospheres or adherent monolayer cultures or primary embryonic or adult cells and prepared as described in Subheadings 2 and 3 above) through a 40 μm cell strainer to remove any undissociated cells or clumps of cells (*see* **Note 9**).
    (a) *Primary embryonic cells* should be at a dilution of 6.5×10^5 cells/mL in Complete Proliferation Medium which will give a final cell plating density of 7,500 cells per 35 mm culture dish in a 25 μL volume.
    (b) *Cells derived from neurosphere cultures or adherent monolayer cultures should be at a dilution of* 2.2×10^5 cells/mL which will give a final cell plating density of 2,500 cells per 35 mm culture dish in a 25 μL volume.
4. Prepare and label the correct number of 35 mm dishes required for the intended experiment. The volumes of reagents listed below are designed for duplicate 35 mm plates.
5. Place a sterile 14 mL tube or the appropriate number of tubes for dispensing the NCFC assay reagents and cells for each test condition in a tube rack.
6. To make the semi-solid collagen NCFC Medium for primary or cultured cells derived from embryonic mouse CNS tissues (allowing duplicate 35 mm culture dishes) add the following components in the given order: 1.7 mL of NeuroCult® NCFC Serum-Free Medium without Cytokines, 0.33 μL of NeuroCult® NSC Proliferation Supplements (Mouse), 6.6 μL of a (rhEGF) stock solution of 10 μg/mL and then 25 μL of the cell suspension either at a concentration of 6.5×10^5 cells/mL (for primary cells) or 2.2×10^5 cells/mL (for cultured cells) (*see* **Note 10**).
7. To make the semi-solid collagen NCFC Medium for primary or cultured cells derived from adult mouse CNS tissues (allowing duplicate 35 mm culture dishes) add the following components in the given order: 1.7 mL of NeuroCult® NCFC Serum-Free Medium without Cytokines, 0.33 μL of NeuroCult® NSC Proliferation Supplements (Mouse), 6.6 μL of rhEGF (stock solution of 10 μg/mL), 3.3 μL of rhFGF (stock solution of 10 μg/mL), 3.3 μL of Heparin (stock solution of 0.2 %) and then 25 μL of the cell suspension either at a concentration of 6.5×10^5 cells/mL (for primary cells) or 2.2×10^5 cells/mL (for cultured cells) (*see* **Note 10**).

8. Mix the medium containing cells (~2.1 mL total) by pipetting with a disposable 2 mL pipette.
9. Using a separate sterile 2 mL pipette, transfer 1.3 mL of cold Collagen Solution to the tube and mix again by pipetting. Using the same 2 mL pipette, remove 1.5 mL of the final culture mixture and dispense this volume into a 35 mm culture dish. Dispense another 1.5 mL in the same manner into a second 35 mm dish. Remove any air bubbles by gently touching bubble with the end of the pipette (*see* **Note 11**).
10. Gently tip each culture dish using a circular motion to allow the mixture in the dishes to spread evenly over the surface.
11. Place the 35 mm culture dishes in a 100 mm petri dish. This petri dish must also contain an open 35 mm culture dish filled with 3 mL of sterile water to maintain optimal humidity during the prolonged incubation period. Replace the lid of the 100 mm Petri dish (*see* **Note 12**).
12. Transfer the plates to an incubator set at 37 °C, 5 % CO_2 and >95 % humidity (*see* **Note 13**).
13. Culture cells for 21 days (differences in colony size can be clearly discerned after 21 days).
14. As cultures are incubated for an extended period of time (21–28 days), cultures should be replenished with Complete Replenishment Medium once a week to avoid depletion of culture media (*see* **Note 14**).
15. Gently (so as not to disrupt gel) add 60 μL of Complete Replenishment Medium into the center of each NCFC dish once every 7 days during the entire NCFC culture incubation (21 days).
16. Cultures should be visually assessed regularly for overall colony growth and morphology using an inverted microscope.

3.3 Categorizing NCFC Colonies and Procedure for Scoring NCFC Colonies

Within 4–7 days of plating in the NCFC Assay, neural stem and progenitor cells begin to proliferate forming small colonies (Fig. 4). By day 14, these small colonies have grown in size and differences can be discerned between colonies. A number of the colonies appear to stop growing after approximately 10–14 days while other colonies continue to expand. By day 21–28, colonies can be classified into four categories: (1) less than 0.5 mm in diameter, (2) 0.5–1 mm in diameter, (3) 1–2 mm in diameter, and (4) 2.0 or >2 mm in diameter. Refer to Fig. 4 for representative examples of different colony sizes.

1. Place an individual 35 mm culture dish on a gridded scoring dish and then place both the culture dish and gridded dish on the dissecting microscope stage.
2. First scan the entire dish using a low power (2.5×–5×) objective lens, noting the relative proximity of the colonies to each

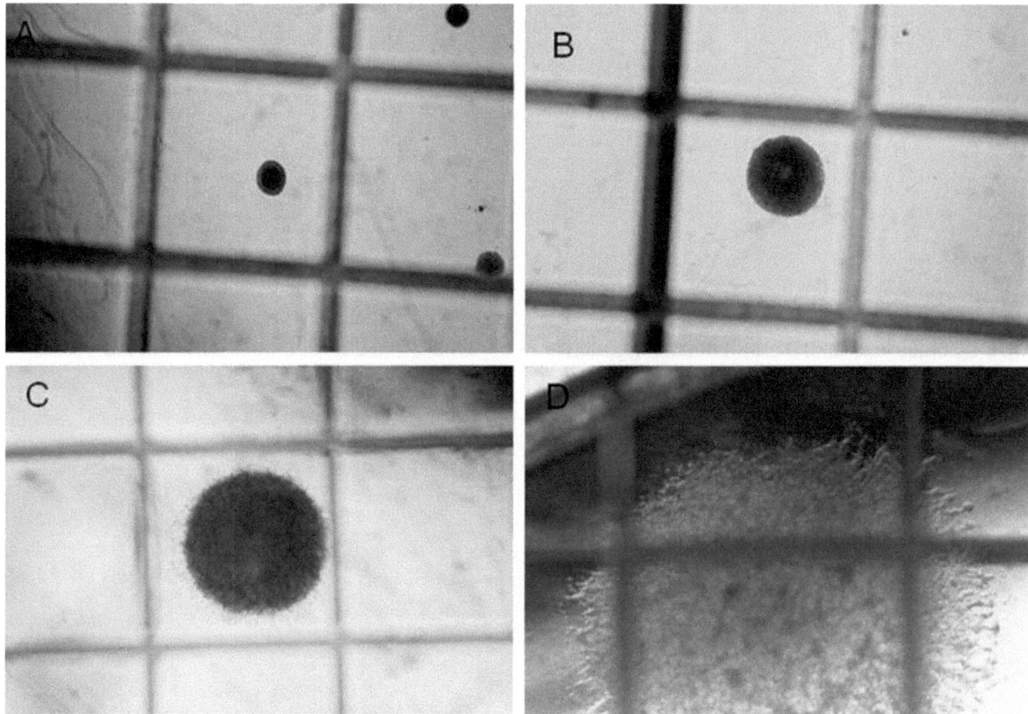

Fig. 4 Representative colony morphologies and size categories derived from embryonic day 14 mouse cortical cells. Colonies were categorized into four size groups: (**a**) less than 0.5 mm in diameter, (**b**) 0.5–1 mm in diameter, (**c**) 1–2 mm in diameter and (**d**) greater than 2 mm in diameter. The grid measures 2 mm × 2 mm

other. Scoring can then be performed with the same lens. Use a higher power (10×) objective to examine colonies in greater detail.

Classify colonies into four categories:

(a) Colonies in the range less than 0.5 mm in diameter (Fig. 5).

(b) Colonies in the range of 0.5–1 mm in diameter (Fig. 6).

(c) Colonies in the range of 1–2 mm in diameter (Fig. 7).

(d) Colonies in the range of 2.0 or >2 mm in diameter (Fig. 8).

3.4 Application of the NCFC Assay for Estimation of NSCs or Neural Progenitors

The following criteria are applied for the quantification of NSCs and progenitor cells from primary embryonic cells or cultured neurospheres derived from embryonic cells:

The original cell that forms a colony 2.0 or >2 mm in diameter is referred to as a Neural Colony Forming Cell–Neural Stem Cell (NCFC–NSC) as this cell has high proliferative potential and multi-lineage potential. Colonies <2 mm contain cells, which lack self-renewal ability and multipotency and are likely produced by a progenitor. For example, using cells from neurospheres generated from E14 striatal cells at passage 2, in the NCFC assay, stem cell frequency (frequency of colonies >2 mm in diameter) was

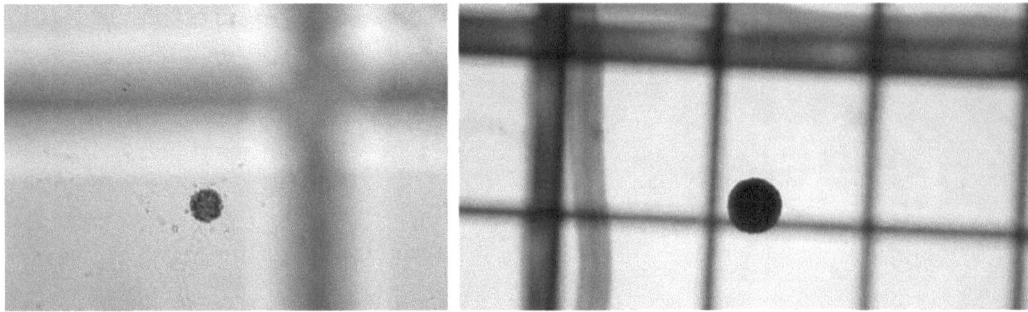

Fig. 5 Colonies less than 0.5 mm in diameter derived from embryonic day 14 mouse cortical cells

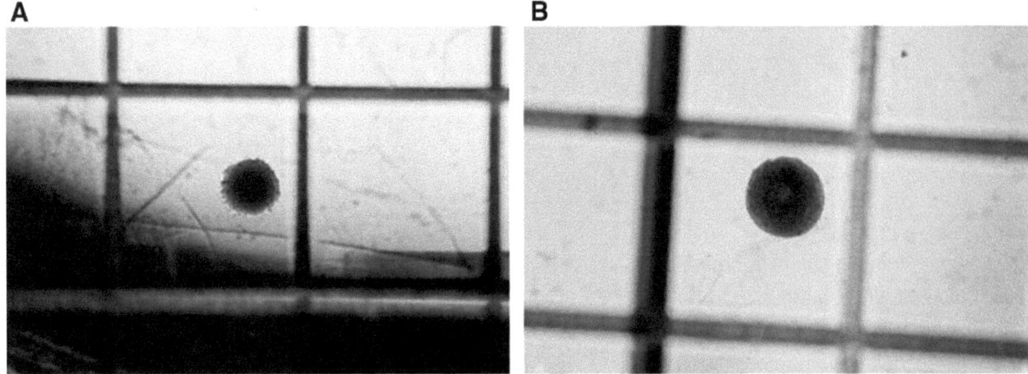

Fig. 6 Colonies within 0.5–1 mm in diameter derived from embryonic day 14 mouse cortical cells. A colony 0.5 mm in diameter (**a**) and a colony 1 mm in diameter (**b**)

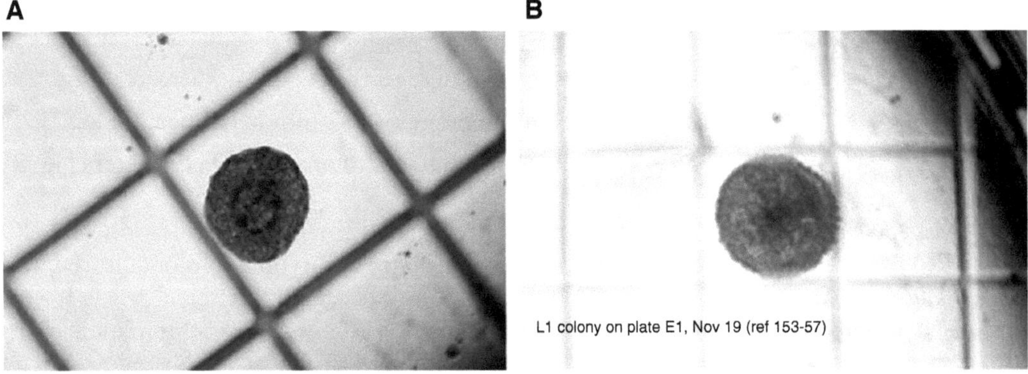

Fig. 7 Colonies 1–2 mm in diameter derived from embryonic day 14 mouse cortical cells. A colony 1 mm in diameter (**a**) and a colony 2 mm in diameter (**b**)

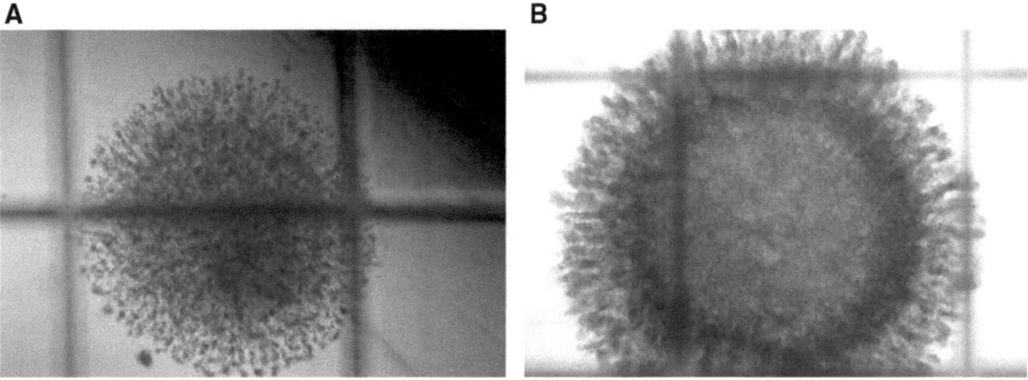

Fig. 8 Colonies >2 mm in diameter derived from embryonic day 14 mouse cortical cells. A colony ~2 mm in diameter (**a**) and a colony 4 mm in diameter (**b**)

estimated to be 0.07 % of total cells, a value comparable to that published using a mathematical model [4, 10]. The NCFC assay can be used for multiple applications including enumerating neural stem cell frequency in vivo [11] and studying their function in mouse model of abnormal brain development [12].

4 Notes

1. The performance of media prepared in the laboratory is highly dependent on the quality and purity of the water and raw materials. If media is prepared in the laboratory, use only tissue culture grade materials and if necessary source various suppliers to determine the best quality reagents as there is significant batch to batch variability in some critical reagents. To avoid variability in media performance, STEMCELL provides optimized and standardized kits for the proliferation of neural cells. An optimized basal medium for the culture of neurospheres from embryonic and adult mouse CNS cells is available, NeuroCult® NSC Basal Media (Mouse) (STEMCELL Technologies Inc.). An optimized 10× proliferation mix for the culture of neurospheres from embryonic and adult mouse CNS cells is available, NeuroCult® NSC Proliferation Supplements (STEMCELL Technologies Inc.). To avoid variability in media performance, STEMCELL provides optimized and standardized kits for the differentiation of neural stem and progenitor cells from the mouse CNS (STEMCELL Technologies Inc.).
2. For adult mouse CNS tissues, process tissues according to standard protocols in the lab or use the NeuroCult® Enzymatic Dissociation Kit for Adult CNS Tissue (STEMCELL

Technologies) and follow the manufacturer instructions. Proceed to **step 6**.

3. To triturate, slightly tilt the tip and press it against the bottom or side of the tube to generate resistance to break up the tissue. The mechanical dissociation of cells by trituration with a fire-polished pipette or disposable plastic tip is known to cause cell death. The use of fire-polished glass Pasteur pipettes is not necessary as disposable plastic tips work well too. Avoid forcing air bubbles into the cell suspensions. Also, it is important to wet the pipette with a small amount of media before sucking the cells into the pipette to reduce the number of cells sticking to the glass or plastic surface.
4. Trituration must be repeated until cell clumps and intact neurospheres are dissociated. Clumps of cells are heavier than single cells, these will settle to the bottom of the tube when left standing for about 5 min. Once the clumps have settled the single cell suspension can be removed to a fresh sterile tube and used for subsequent cultures, leaving the clumps at the bottom of the tube. Repeat this procedure of dissociating the remaining undissociated clumps, allowing clumps to settle at the bottom and removing the supernatant containing the single cell suspension into the tube containing the rest of single cell suspension until the majority of clumps have been dissociated.
5. To set up adherent monolayer cultures, coating the surface of the tissue culture vessels with laminin or poly-D-lysine (PDL)/laminin is required. Prepare 0.1 mg/mL PDL solution by dissolving 5 mg PDL (Sigma) in 50 mL sterile water. Prepare 10 μg/mL laminin by diluting laminin (Sigma) with sterile water. To coat tissue culture vessels, add 1 mL of substrate (0.1 mg/mL PDL or 10 μg/mL laminin) to each well in a 6-well tissue culture dish or 3 mL per T-25 cm^2 flask. Incubate for 2 h at 37 °C or overnight at 2–8 °C. When ready to plate cells, remove substrate and wash tissue culture vessels with sterile PBS. Do not let the coated plates completely dry.
6. Other protocols are available to dissociate neurospheres to produce a single cell suspension. For example, the NeuroCult® Chemical Dissociation Kit (STEMCELL Technologies, Inc.) offers a nonmechanical, nonenzymatic alternate procedure for dissociating neurospheres derived from embryonic and adult mouse CNS cells which yields greater cell viabilities compared to the trituration method. Accutase™ (STEMCELL Technologies Inc.) can also be used for dissociating neurospheres derived from embryonic and adult mouse CNS without adversely affecting stem cell function.
7. The volume of the pipettor is set lower that the total volume of the cell suspension to avoid expulsion of all the liquid and introduction of air bubbles into the cell suspension.

8. Keep the collagen on ice throughout the culture set up to prevent the collagen from gelling at higher temperatures.
9. The NCFC assay is based on the formation of clonally derived colonies from a single colony forming cell. This is possible by ensuring that the start cell suspension is provided as a single cell suspension .
10. The cell plating density was determined by titration experiments and determining the linearity ranges for primary cells isolated from normal embryos of pregnant CD1 albino mice or cells from neurospheres derived from normal E14 CD1 albino mice cortices and/or striata cultured for two passages. It may be necessary to perform titration curves within the range of 5,000–50,000 cells per dish for primary cells or 1,000–5,000 cells per dish for cultured cells when different species and transgenic animals are used. The cloning efficiency may be different for different starting cell populations. Ensure that initial concentration of the cells is adjusted so that only a total volume of 25 μL of cells is always added to the medium mixture described in Subheading 3.2 to maintain accurate media concentrations.
11. The collagen solution is always added last. If multiple tubes are being set up, add cells to a single tube then add collagen and plate cells. Do not let cells sit in NCFC medium for an extended period of time before plating. The collagen starts to gel within several minutes following the addition to the cell suspension. If more than one tube is being setup, collagen should be added to the first tube only, and the contents dispensed into dishes before proceeding to the next tube.
12. If many dishes are used, these dishes can also be placed in a covered 245 mm square bioassay dish with two or three open 35 mm culture dishes containing sterile water.
13. Gel formation will occur within approximately 1 h. It is important not to disturb the cultures during this time.
14. Make up 10 mL of Complete Replenishment Medium by mixing 9 mL of Basal Medium and 1 mL of 10× Proliferation Mix. To this, add 500 μL of the 10 μg/mL stock solution of hEGF. The Complete Replenishment Medium contains 1:10 Basal and 10× Proliferation Mix and 0.5 μg/mL of hEGF. Store the prepared media at 4 °C for up to the 3 weeks of replenishing.

Acknowledgments

We would like to acknowledge our collaborators Dr. Brent Reynolds and Loic Deleyrolle, for technical and scientific help and discussions.

References

1. Reynolds B, Weiss S (1992) Generation of neurons and astrocytes from isolated cells of the adult mammalian central nervous system. Science 255:1701–1710
2. Reynolds BA, Tetzlaff W, Weiss S (1992) A multipotent EGF-responsive striatal embryonic progenitor cell produces neurons and astrocytes. J Neurosci 12:4565–4574
3. Reynolds BA, Weiss S (1996) Clonal and population analyses demonstrate that an EGF-responsive mammalian embryonic CNS precursor is a stem cell. Dev Biol 175:1–13
4. Reynolds BA, Rietze RL (2005) Neural stem cells and neurospheres – re-evaluating the relationship. Nat Methods 2:333–336
5. Rietze RL, Reynolds BA (2006) Neural stem cell isolation and characterization. Methods Enzymol 419:3–23
6. Singec I et al (2006) Defining the actual sensitivity and specificity of the neurosphere assay in stem cell biology. Nat Methods 3:801–806
7. Louis SA et al (2008) Enumeration of neural stem and progenitor cells in the neural colony-forming cell assay. Stem Cells 26(4):988–96
8. Conti L et al (2005) Niche-independent symmetrical self-renewal of a mammalian tissue stem cell. PLoS Biol 3:e283
9. Conti L, Cattaneo E (2010) Neural stem cell systems: physiological players or in vitro entities? Nat Rev Neurosci 11:176–187
10. Deleyrolle LP et al (2011) Determination of somatic and cancer stem cell self-renewing symmetric division rate using sphere assays. PLoS One 6:e15844
11. Bull ND, Bartlett PF (2005) The adult mouse hippocampal progenitor is neurogenic but not a stem cell. J Neurosci 25(47): 10815–10821
12. Siji-Felice K et al (2008) Fanconi DNA repair pathway is required for survival and long-term maintenance of neural progenitors. EMBO J 27(5):770–781

Part III

Purification of Stem and Progenitor Cells

Chapter 12

Flow Cytometry of Neural Cells

Geoffrey W. Osborne

Abstract

Flow cytometry is an advanced group of techniques for counting and quantifying microscopic particles such as cells, chromosomes, or functionalized beads. These approaches employ sophisticated optical and fluidic components to detect scattered light and fluorescent signals from cells as they sequentially pass an interrogation point. Cytometry plays a crucial role in the diagnosis of immunological disorders and cancers, and is a mainstay technique in basic research settings such as hematology, cell biology, and biomolecular screening. However, in spite of the breadth of applications spanning many fields, flow cytometry in neuroscience has been largely unexploited and has seen only a steady increase in interest until recent years. This is rather surprising as the potential of flow cytometry in neuroscience applications was recognized in the early 1980s as the technology was evolving.

Keywords Flow cytometry, Neural cells, Fluorescence, Imaging, Antibody, Fluorochrome, Isotype

1 Introduction

Flow cytometry is an advanced group of techniques for counting and quantifying microscopic particles such as cells, chromosomes, or functionalized beads. These approaches employ sophisticated optical and fluidic components to detect scattered light and fluorescent signals from cells as they sequentially pass an interrogation point. Cytometry plays a crucial role in the diagnosis of immunological disorders and cancers, and is a mainstay technique in basic research settings such as hematology, cell biology, and biomolecular screening. However, in spite of the breadth of applications spanning many fields, flow cytometry in neuroscience has been largely unexploited and has seen only a steady increase in interest until recent years. This is rather surprising as the potential of flow cytometry in neuroscience applications was recognized in the early 1980s as the technology was evolving.

Brent A. Reynolds and Loic P. Deleyrolle (eds.), *Neural Progenitor Cells: Methods and Protocols*, Methods in Molecular Biology, vol. 1059, DOI 10.1007/978-1-62703-574-3_12, © Springer Science+Business Media New York 2013

2 Overview of the Technology

In order to fully capture the potential for flow cytometry with neural cells, a thorough understanding of the strengths and weaknesses of the technology need to be grasped. It should, however, be recognized at the outset that most of the limitations are related to the lack of suitable probes with appropriate specificities and good quality single cell preparations. If these factors can be suitably addressed then the potential for flow cytometry assays is enormous.

Cells of neural origin need to be made into single cell suspensions before being suitable to pass through the instrument. The cells container is pressurized and the cells are surrounded by a sheath fluid and conditions of laminar flow established so that cells line up one after the other in a single file to pass through a laser interrogation point. The interrogation point consists of an elliptical beam spot generated by one or more lasers where optical signals are generated, collected, and then converted to digital outputs for computer display and storage. The collected signals are either intrinsic scattered light or fluorescence signals, or fluorescence signals generated by labels either in or on the cells. In normal circumstances, the fluorescence is associated with labelled antibodies that have an affinity to cell surface receptors, and one assumes some specificity in binding, or the fluorescence could be from a probe associated with an internal physiological indicator such as calcium (INDO-1 probe), mitochondrial membrane potential (JC-1 probe) [1], or lipophilic tracers dyes [2]. Modern cytometers can measure multiple parameters simultaneously and this combined with information from surface and internal reporters measured simultaneously provide many options to elucidate the way interactions occur at the cellular level.

The technology is more valuable when coupled with electrostatic cell sorting which was first developed in 1965 [3]. The theoretical and practical aspects are covered extensively in other works [4, 5]; however a précis here may aid the reader. In simple terms when cells are to be sorted, the previously mentioned stream of cells is vibrated so that the stream accurately and reproducibly breaks into drops, and, at an precisely timed interval after the sample interrogation point and, as a drop containing a particle of interest is about to break off, the stream is electrostatically charged and then deflected into a collection vessel. The collected cells are then available either for clonal analysis following sorting into multi-well plates [6], or collected into dishes or tubes containing culture medium for further experimentation. These experimental approaches and the tips for successful flow cytometry experiments will be discussed in the remainder of this chapter.

3 Uses of Flow Cytometry in the Analysis of Neural Cells

When considering whether to utilize flow cytometry as a research tool, it is worthwhile clearly defining what is the purpose of the experiment, the source of cells you plan to use, how you are going to dissociate them, what method or methods you will use for identifying the cells of interest.

For the majority of the work involving neural cells, a dissociation step is almost a mandatory requirement, as unlike haematopoietic samples where cells are in suspension, cells of neural origin are either from solid tissue or are tissue derived and maintained or expanded in culture. This was recognized [7] as soon as commercial instrumentation began to become widely available, with early papers detailing the separation of neurons and oligodendrocytes based on intrinsic autofluorescence properties from dissociated tissue. The primary goal of the dissociation step should be to yield a suspension of single cells.

A single cell suspension is a requirement for flow cytometry, or at the very least clumps of cells that are small enough to be able pass through the instrument without clogging it is required. There are two main methods for tissue dissociation, enzymatic digestion, and mechanical disruption, that are normally used consecutively. These methods are covered in some detail in other chapters of this book and are therefore not dealt with here. However the effect different methods of dissociation can have on flow cytometric results needs to be covered.

Consider the panels of Fig. 1 showing cultured cortical neurons expressing green fluorescent protein (GFP) growing in an adherent manner in culture dishes prior to trypsin treatment to remove cells from the plastic. Axons are clearly evident projecting from the cell body of the cell both in the bright field and fluorescent images (Fig. 1a). Post trypsinization, the cells were resuspended in phosphate-buffered saline and run on both an imaging flow cytometer (panel b) and normal flow cytometer (panel c).

As is evident in Fig. 1b, the cells have been dramatically affected by the trypsin treatment with a variety of cell shapes evident. Clumps of cells, cellular debris, and relatively intact cells (red-boxed enlarged section) that have maintained axons are all present. Given the bright field images show such a great diversity of cell morphologies, it therefore follows that this diversity will be represented in the way light is scattered from the cells or particles as they pass through the focused laser beam. Typical resulting data are shown in Fig. 1c right panel with a large range of forward and side scattered light signals evident. The preparation also highlights a characteristic of neural tissue preparations and that is the presence of many small particles and large amounts of cellular debris that tend to pile up in the lower corner of the flow cytometry bivariate display of scattered light parameters. It is important to recognize that this is a trap for those unfamiliar with neural flow cytometry as the large amounts of debris can sometimes “swamp” the presence

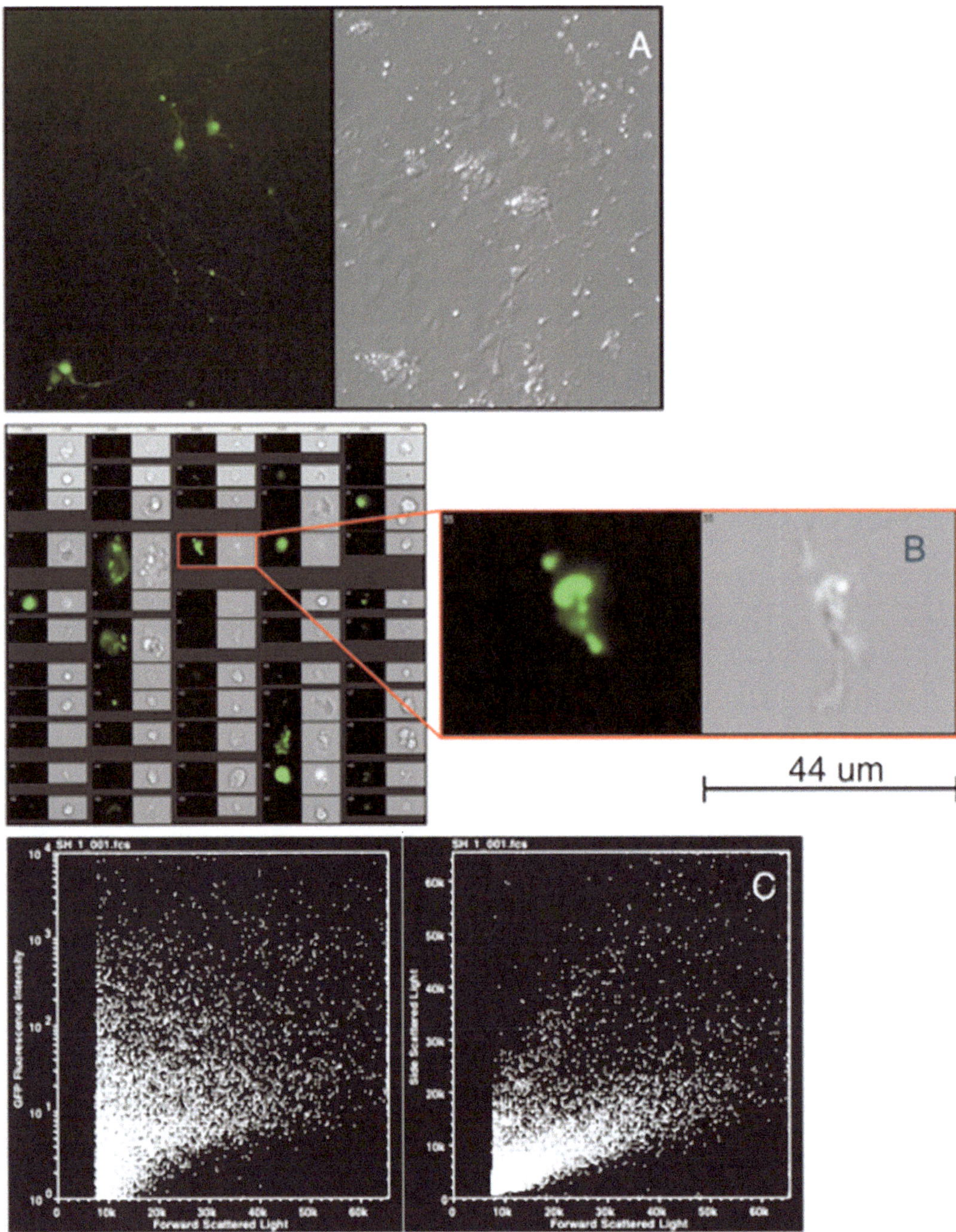

Fig. 1 Mouse cortical neurons grown in culture. (**a**) Fluorescence (Green) and corresponding bright field images of neurons adhered to plastic prior to trypsinization. Note the relative length of axons and that most soma have associated processes. (**b**) AMNIS imaging cytometry showing morphological changes from the cells shown in *panel* (**a**) resulting from trypsin treatment, with most cells having axons either removed, rounded up or processes retracted. (**c**) Flow cytometry bivariate displays of (*lower left*) Forward Scattered Light versus relative fluorescence intensity GFP from the same sample preparation shown in *panel* (**b**). Scattered light signals (*lower right*) reflect the diversity present within the preparation with the majority of debris falling in the plots' lower left corner and the population with higher side scattered light and lower forward scatter representing dead or apoptotic cells

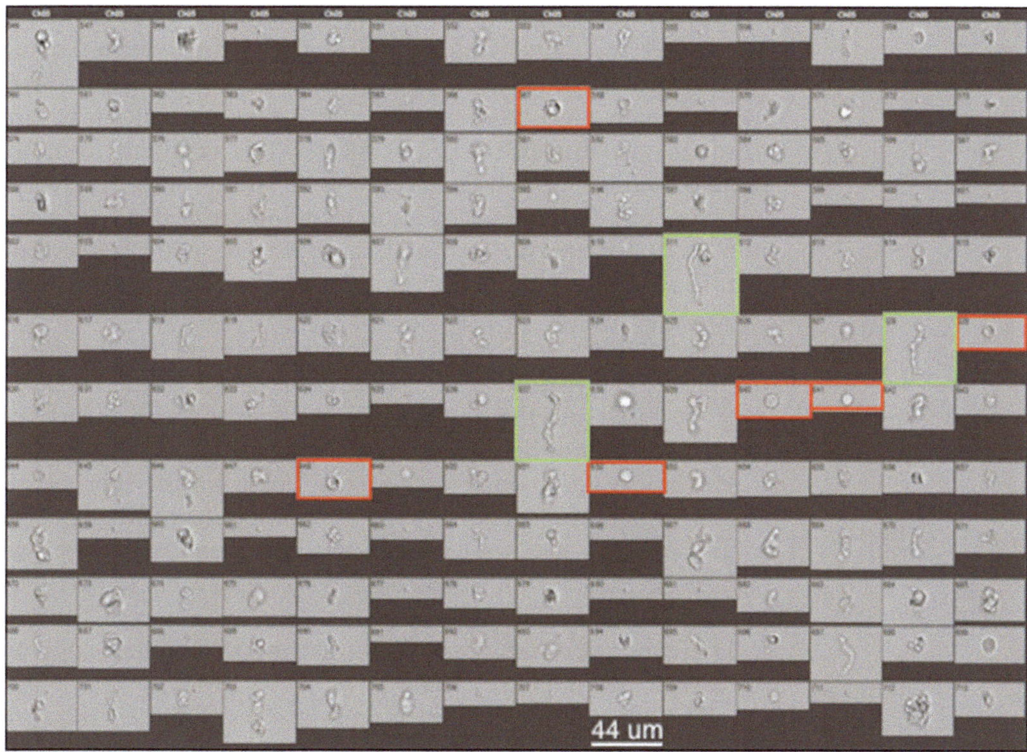

Fig. 2 Mouse hippocampal cells dissociated using Papain. Imaging cytometry bright field images demonstrate a typical flow cytometry cell preparation resulting from papain dissociation. Single cells (*red boxes*) are evident, as are intact large processes (*green boxes*) no longer associated with soma. Additional images given an indication of the relative numbers of round cells expected compared to cellular debris. The cellular debris represents the majority of the particles within a typical primary tissue preparation. Scale bar/image width = 44 μm

of smaller intact cells. In order to remove the debris, the temptation may be to increase the instruments system threshold, the minimal level of an instrument trigger parameter that needs to be satisfied in order to be counted by the system as a particle of interest. If the threshold is set too high then the very cells that are of interest, can be excluded from analysis. Conversely, if a priori knowledge indicates that only larger particles are of interest, an increase in the system threshold to exclude more debris and small cells may indeed be the correct approach, so some guidelines for instrumentation adjustment will always prove to be beneficial.

The other common method of dissociating cells for flow cytometry assays is the use of the cysteine protease Papain. Papain has been shown to be effective a dissociating certain types of cortical neurons [8] while maintaining cell surface functionality. Experience indicates that a "trial and error" approach is commonly required to obtain the correct combination of conditions to optimally tease out the cells of interest from a particular tissue type and the affect of papain on mouse hippocampal preparations used for flow cytometry are apparent in Fig. 2. Imaging cytometry results shown in Fig. 2 highlight the problem of cellular debris that is characteristic of working with dissociated primary neural tissue. Relative to the small number of single cells, which are normally required for flow cytometry, the majority of particles passing through the instrument are cellular debris resulting from the required sample preparation protocols. Please consider this in your experimental design by using fluorescent indicators that stain either the nucleus or the cytoplasm and can be used to set the flow cytometer so that only stained particles are processed, analyzed, or sorted by the instrument.

In addition to straight trypsin and papain treatments to dissociate cells, there are a number of commercial products available that are based on either of these agents combined with proprietary solutions that reportedly provide excellent results. These solutions need to be assessed on a "case-by-case" basis for the appropriateness to a particular experiment, however regardless of the dissociation method in addition to single cells, debris and cell clumps will always be present. On instruments with changeable sample nozzles, it is therefore recommended to use a minimum of a 90 μm diameter nozzle as this decreases the likelihood of system clogs due to cellular debris and cell clumps.

The separation of the main types of cells from neural central nervous system (CNS) tissue is now documented [9–11] and readers are advised to consult these works if this is your area of interest. Similarly, the use of magnetic separation methods for the preparation of cell samples [12] may be beneficial as it possible to process large numbers of cells using antibodies conjugated to magnetic particles. The number of cells enriched in this manner can far exceed that which is possible by flow cytometric sorting in a similar time. Care needs to be taken that the correct dissociation treatment is used so that the antigen is not stripped, with certain antigens for example A2B5 for glial restricted precursors [13] preferring papain and others PSA-NCAM requiring trypsin.

However be aware that by binding an antibody to the cell may trigger an unwanted cascade of cellular events, that may or may not be detectable and which is avoided when using density centrifugation-based methods. In addition there is an "all or none" type separation occurring that is not the case when selecting cells using flow cytometry methods, where it is possible to separate out groups of cells which bind to receptors on the cell surface at varying densities. Thus for example by flow cytometry one could select only the cells with high, intermediate, or low antibody binding levels and simultaneously collect each population for further experimentation. Importantly, density centrifugation steps followed by sorting have been shown to be beneficial is improving cell recovery in neurogenesis studies [14]. Regardless of the collection strategy the reader is encouraged to think about the possible implications of the separation strategy well before undertaking an experiments.

Once cells are available as single cell suspensions, the power of the flow cytometry can be leveraged in the types of assays that can be performed. As many assays rely on binding antibodies, some background considerations are important.

It is always preferable to use a monoclonal antibody, and preferable the Fab_2 fragment of the antibody wherever possible to ensure the greatest specificity of binding. Monoclonal antibodies are very epitope specific and probably conformation specific as well. However, where polyclonal and not monoclonal antibodies are available then these antibodies can be used with the anticipation that the specificity will be lower than with a monoclonal equivalent. Increasingly engineered antibodies generated by methods such as bacteriophage technology [15] are yielding new antibodies appropriate for use with neural cells. However, unfortunately many of the available antibodies for neuroscience have not been optimized for flow cytometry but for western blot (WB) analysis or immunohistochemistry applications. The issue is that the WB reagents are normally raised against a peptide, often chosen based on solubility and sequence diversity from endogenous homologous protein, present in the host organism. Practically this means a lot of the epitopes are buried inside the intact protein when it is properly folded on the cell membrane, and therefore the antibody that works for WB may not be suitable for flow cytometry. However experience indicates that the available antibody should be trialled regardless of the stated assay specificity, as some antibodies will prove to be suitable for neural cells.

Antibodies that are specified for immunoprecipitation protocol (IP) and immunohistochemistry (IHC) may also be usable and should be tested for binding for flow cytometry. It is important that some additional testing be performed to confirm that the IP- or IHC-specified antibodies actually bind to the expected site by either competing for the antigen binding site with another known antibody which binds to the target or by a microscopy-based assessment. In the case of IHC, many neuroscience antibodies are specified to function in this format when tissues are fixed in the form of histology sections and bind to the appropriate binding site when the cellular proteins are cross linked and the cell permeabilized to permit antibody egress.

Many flow cytometry experiments require the use of cells that have undergone a fixation process. The inclusion of a fixation step may be deemed necessary with the realization that significant periods of time are involved from staining samples until they can be analyzed, or fixation may be an essential part of the staining protocol. In the latter case, fixation is usually combined with a detergent treatment to facilitate antibody entry into the cell. A variety of combinations of fixations and detergent treatment have been optimized for flow cytometry [16, 17] and it is worth referring to the body of work in this area before undertaking new experiments.

Since experiments often involve the use of fluorochrome-labeled antibodies, appropriate controls can prove valuable, however just what is the appropriate control can be difficult, if not impossible to establish. Conceptually, isotype control antibodies have no relevant specificity help to distinguish nonspecific “background” staining from specific antibody staining, and binding is reported by a fluorochrome conjugated to the isotype control antibody. However there is a problem using isotype controls which is often overlooked or unrecognized even by experienced cytometrists that relates to ratio of the number of fluorochrome molecules bound to each antibody protein. This can be difficult to control when performing the fluorochrome conjugations, and for example one manufacture may have a conjugate with three fluorescent molecules per antibody, while another manufacturer has four fluorescent molecules per antibody. The instrumentation measures the light collected from the fluorescent molecules and therefore reports different amounts of nonspecific binding for the isotype. Compounding the problem is that the specific antibody may also have a different fluorochrome/protein (F/P) ratio to the isotype control.

The interpretation of results when using isotype controls can also be problematic [18]. Take the scenario where some binding of the isotype control antibody occurs and then binding of the specific antibody also occurs perhaps at a higher level. It is possible to subtract the percentage of cells of binding of the isotype control from the percentage of cells binding the specific antibody, although this may be a flawed strategy as another different isotype may show different levels of background binding. In addition, when multiple antibodies are used simultaneously with the same F/P conjugates, one may have the fluorochrome conjugated at a critical “background” binding site, while another does not, and this can dramatically affect the “stickiness” or binding to Fc receptors and amount of background detected.

In addition, considered the problem of what would be an appropriate concentration to use for any selected isotype control? For specific antibodies, one can choose the “saturating” concentration, however for the isotype control antibody there is no such thing as saturation; the more antibody you use, the more background you obtain. If we use an isotype control we could match the concentration to that of the test antibody, on the possibly false assumption that the proper minimal saturating concentration for one would mimic another. Coupled with this, variations in intrinsic fluorescence that the different cell populations possess, due in part to NADPH levels [19] and flavin content, can vary the background considerably. With these factors in mind, control considerations for flow cytometry experiments need to be approached with care and further guides for the reader are available [20] which provide additional points for consideration.

The range of flow cytometry assays available to the researcher is now extensive, and it is beyond the scope of this chapter to capture the breadth of neuroscience applications that utilize the technology. Rather, the reader is recommended to identify the cell type of interest and then "home in" on the literature for the area of expected involvement, be it brain, CNS, or peripheral nervous system. An example of this would be studies of microglia, where ones interest may lie in their involvement in stroke [21, 22], in the CNS [23–25] or the involvement in immune responses to viral challenge or age-related decline. Within each of these areas there are now bodies of literature that can be accessed.

In summary, this chapter has attempted to provide the reader with a flavor of the range of applications that can be undertaken using flow cytometric techniques, while providing general information to keep in mind when dissociating and staining samples. New and novel methods continue to emerge, such as the tracking of protein aggregation and mislocalization in a mouse cell line model of Huntington's disease [26] that demonstrate of the utility of flow cytometry in the context of neuroscience research. It is hoped that these new methods, along with more well established methods in the areas of stem and progenitor cell biology, will encourage greater utilization of this powerful technology and benefit other research efforts involving neural samples.

References

1. RIN M et al (1988) Flow cytometric analysis of membrane potential in embryonic rat spinal cord cells. J Neurosci Methods 22:203
2. Li X et al (2003) Labeling Schwann cells with CFSE–an in vitro and in vivo study. J Neurosci Methods 125:83
3. Fulwyler MJ (1965) An electronic particle separator with potential biological application. Science 150:371
4. Herzenberg LA et al (2002) The history and future of the fluorescence activated cell sorter and flow cytometry: a view from Stanford. Clin Chem 48:1819
5. Ibrahim SF, van den Engh G (2007) Flow cytometry and cell sorting. Adv Biochem Eng Biotechnol 106:19
6. Battye F, Light A, Tarlinton D (2000) Single cell sorting and cloning. J Immunol Methods 243:25
7. Meyer R, Zaruba M, McKhann G (1980) Flow cytometry of isolated cells from the brain. Anal Quant Cytol 2:66
8. Huettner J, Baughman R (1986) Primary culture of identified neurons from the visual cortex of postnatal rats. J Neurosci 6: 3044
9. Dobrenis K (1998) Microglia in cell culture and in transplantation therapy for central nervous system disease. Methods 16:320
10. Moussaud S, Draheim HJ (2010) A new method to isolate microglia from adult mice and culture them for an extended period of time. J Neurosci Methods 187:243
11. Guez-Barber D et al (2012) FACS purification of immunolabeled cell types from adult rat brain. J Neurosci Methods 203:10
12. Marek R et al (2008) Magnetic cell sorting: a fast and effective method of concurrent isolation of high purity viable astrocytes and microglia from neonatal mouse brain tissue. J Neurosci Methods 175:108
13. Rao MS, Noble M, Mayer-Pröschel M (1998) A tripotential glial precursor cell is present in the developing spinal cord. Proc Natl Acad Sci 95:3996
14. Spoelgen R et al (2011) A novel flow cytometry-based technique to measure adult neurogenesis in the brain. J Neurochem 119:165
15. Hoogenboom H (2005) Selecting and screening recombinant antibody libraries. Nat Biotechnol 23:1105
16. Chow S et al (2005) Whole blood fixation and permeabilization protocol with red blood cell lysis for flow cytometry of intracellular phosphorylated epitopes in leukocyte subpopulations. Cytometry A 67:4
17. Hedley DW, Chow S, Shankey TV (2011) Cytometry of intracellular signaling: from labo-

ratory bench to clinical application. Methods Cell Biol 103:203
18. O'Gorman MRG, Thomas J (1999) Isotype controls—time to let go? Cytometry 38:78
19. Tzur A et al (2011) Optimizing optical flow cytometry for cell volume-based sorting and analysis. PLoS One 6:e16053
20. Maecker HT, Trotter J (2006) Flow cytometry controls, instrument setup, and the determination of positivity. Cytometry A 69A:1037
21. Campanella M et al (2002) Flow cytometric analysis of inflammatory cells in ischemic rat brain. Stroke 33:586
22. Gelderblom M et al (2009) Temporal and spatial dynamics of cerebral immune cell accumulation in stroke. Stroke 40:1849
23. White CA, McCombe PA, Pender MP (1998) Microglia are more susceptible than macrophages to apoptosis in the central nervous system in experimental autoimmune encephalomyelitis through a mechanism not involving Fas (CD95). Int Immunol 10:935
24. Piccio L et al (2008) Identification of soluble TREM-2 in the cerebrospinal fluid and its association with multiple sclerosis and CNS inflammation. Brain 131:3081
25. Liu Y et al (2005) LPS receptor (CD14): a receptor for phagocytosis of Alzheimer's amyloid peptide. Brain 128:1778
26. Ramdzan YM et al (2012) Tracking protein aggregation and mislocalization in cells with flow cytometry. Nat Methods 9:467

Part IV

Transplantation of Neural Progenitor Cells: Preclinical Models

Chapter 13

Neonatal Transplant in Hypoxic Injury

Tong Zheng and Michael D. Weiss

Abstract

Hypoxic–ischemic encephalopathy in neonates often causes long-term disabilities. Stem cell therapy may be a successful treatment for HIE. Neurogenic astrocytes with characteristics of neural stem cells (NSCs) can be cultured as adherent monolayers. Following reintroduction into the NSC niche of both neonatal and adult hosts, these astrocytes can be induced to generate neuronal progeny in vivo. Thus, neurogenic astrocytes represent promising candidates for cell replacement therapy in HIE. Such an approach requires optimized cell cultivation protocols as well as extensive testing of donor cells to assess their capacity for engraftment, survival, and integration in the HIE animal models. In this chapter, we describe methods of generating the HIE model, generating and culturing monolayer neurogenic astrocytes, and transplanting these cells into HIE animal models.

Key words Transplantation, HIE, Neural stem cells, Neurogenic astrocytes, Subependymal zone

1 Introduction

Hypoxic–ischemic encephalopathy (HIE) is the brain manifestation of systemic asphyxia [1] and occurs in about 20 out of 1,000 full-term infants and in nearly 60 % of very low birth weight (premature) newborns [2–4]. Results from four multicenter trials [5–8] indicate that hypothermia diminishes brain injury in some neonates with moderate to severe HIE. Currently, hypothermia is considered a state-of-the-art therapy, bordering on becoming a standard of care, for neonates with HIE. Meta-analysis suggests the therapy will prevent long-term neurologic deficits in one neonate for every eight treated [9]. Since hypothermia does not help every neonate with HIE, researchers are searching for therapies to use in combination with hypothermia [10, 11]. While neuroprotection may reduce ongoing or escalating damage, neonates with moderate to severe HIE may benefit from cellular replacement to repair already-damaged regions. The neonate may be an ideal candidate for a cellular-based therapy due to the increased plasticity at that early developmental stage. In studies, neural stem cells (NSCs)

Brent A. Reynolds and Loic P. Deleyrolle (eds.), *Neural Progenitor Cells: Methods and Protocols*, Methods in Molecular Biology, vol. 1059, DOI 10.1007/978-1-62703-574-3_13, © Springer Science+Business Media New York 2013

grafted into neonatal rats developed into functional pyramidal neurons and integrated into host cortical circuitry [12]. In hopes of repopulating or minimizing the endogenous loss of cells, researchers have injected a variety of cell types, including NSCs, fetal cortex, human umbilical cord blood, bone marrow-derived multipotent adult progenitor cells, and multipotent astrocytic stem cells (reviewed in ref. 13). Our group explored regenerative medicine as a possible therapy for HIE by employing the Rice-Vannucci model in both rats and mice [14]. This small-animal model of hypoxic–ischemic injury has been extensively used to test emerging neuroprotective therapies for neonates. When compared to large-animal models, the Rice-Vannucci model is easier to use, produces faster results, and can be performed on larger sample sizes in a shorter period of time [15, 16]. As a result, this model is an ideal platform to begin studying regenerative medicine.

The adult mammalian brain harbors within the periventricular subependymal zone (SEZ) a pool of multipotent NSCs. These NSCs have the capacity for multi-lineage differentiation. Researchers are interested in utilizing NSCs in transplantation approaches. Since NSCs represent only a small population of SEZ cells, in vitro expansion is required to generate a sufficient number of cells for transplantation. Neurogenic astrocytes derived from the SEZ respond to intrinsic environmental cues by anatomically integrating into the host nervous system when transplanted into the brain [17]. These astrocytes may provide promising donor cells for therapeutic grafting to treat HIE.

In this chapter, we discuss the surgical technique used to produce the hypoxic–ischemic injury, the timing of cell transplants, the preparation and culture of the neurogenic astrocytes derived from the subventricular zone, and the transplant process.

2 Materials

2.1 Hypoxic–Ischemic Injury

1. Surgical area that can be easily cleaned between pups.
2. Cautery device: I-STAT® Low Temperature Cautery, Fine Tip (Medtronic, Minneapolis, MN).
3. Surgical instruments: Noyes micro dissecting spring scissors, straight (Roboz Surgical Instrument Co., Gaithersburg, MD, cat# RS-5676), Bonn micro dissecting scissors, straight (Roboz, cat# RS-5840), extra-fine curved Graefe forceps (Fine Science Tools®, Foster City, CA; cat# 11151-10), and Graefe serrated curved forceps (FST®, cat# 11052-10), straight Graefe forceps (FST cat# 11050-10).
4. Anesthesia delivery device: Ohmeda Isotec 4 Isoflurane Vaporizer (Eagle Eye Anesthesia Jacksonville, FL).

5. Rodent nose cone and induction chamber.
6. Dissecting Microscope: ZoomMaster 65 (Braintree Scientific, Inc., Braintree, MA).
7. Skin Glue: Vetbond™ (3 M™, St. Paul, MN).
8. Cotton tip applicator and gauze.
9. Acu-Rite® Sensor with a Resistance Temperature Detector (RTD).
10. Warming pads: Deltaphase® Isothermal Pad (Braintree Scientific, Inc., Braintree, MA).
11. Alcohol swabs.
12. Povidine–iodine.
13. 8 % oxygen, balance with nitrogen.
14. Isoflurane (Webster Veterinary, Devens, MA, cat# 07-8366551).
15. Buprenorphine, 0.01–0.05 mg/kg IM (Webster Veterinary, Devens, MA, cat# 07-850-2280).
16. Neosporin® antibiotic ointment (Johnson & Johnson Consumer Companies, Inc., New Brunswick, NJ).
17. Modular incubator chamber (Billups-Rothenberg, Inc., Del Mar, CA, cat# MIC-101).

2.2 Preparation of Neurogenic Astrocytes for Transplant

1. Base medium: Dulbecco's Modified Eagle's Medium: Nutrient Mixture F-12 (DMEM/F12 medium, Invitrogen™, Grand Island, NY, cat# 11320) containing antibiotic/antimycotic.
2. N2 Supplement: 500× Stock Solution (Invitrogen™, Grand Island, NY, cat# 17502-048).
3. Bovine Pituitary Extract (BPE, Invitrogen, Grand Island, NY, cat# 13028-014).
4. Fetal bovine serum (Atlanta Biologicals®, Lawrenceville, GA, cat# S11050).
5. Growth medium: Base medium containing N2 supplement, BPE (20 μg/mL), 5 % FBS, epidermal growth factor (EGF, 20 ng/mL, Sigma®, St. Louis, MO, cat# E9644), and basic fibroblast growth factor (bFGF, 10 ng/mL, Sigma®, St. Louis, MO, cat# F0291).
6. 0.25 % Trypsin/EDTA solution (Atlanta Biologicals®, Lawrenceville, GA, cat# B81310).
7. Fire-polished Pasteur pipettes: prepare medium- and narrow-bore sets by briefly exposing the tip of the pipette to the flame of a Bunsen burner to narrow the lumen.
8. Conical centrifuge polypropylene tubes, 15 mL (Midwest Scientific, Valley Park, MO, cat# TP91015).

9. Growth factor stock solution: Cultures require supplementation with 20 ng/mL of EGF and 10 ng/mL of bFGF every 2–3 days. Since each culture will contain approximately 12 mL of medium, supplement with 300 μL aliquots of 40× stock (8,000 ng of EGF and 4,000 ng of bFGF in 10 mL of DMEM/F12).
10. Dulbecco's PBS (dPBS, Invitrogen, Grand Island, NY, cat# 14190): containing antibiotic/antimycotic.
11. T-75 tissue culture flasks (Midwest Scientific™, Valley Park, MO, cat# TP90076).

2.3 Transplantation of Neurogenic Astrocytes

1. Isofluorane (Webster Veterinary, Devens, MA, cat# 07-8366551) for anesthesia.
2. Modeling clay mold for holding the pups in place.
3. Small container for mouse pups.
4. Alcohol wipes.
5. Cotton tip applicators.
6. Micromanipulator for Hamilton syringe (Stoelting Company, Wood Dale, IL, cat# 55133 base cat#55116).
7. Hamilton syringe (Hamilton Company, Reno, NV, cat#641-01 5 μL Model 85RN SYR).
8. 33 G Hamilton Needle (Hamilton Company, Reno, NV, cat # 7803-05, Small Hub RN NDL. Custom length 1 in., Point style 4).
9. Warming pad with temperature controlled at 37 °C to warm pups after surgery (Braintree Scientific, Inc., Braintree, MA, cat# ASS7T).

3 Methods

3.1 Hypoxic–Ischemic Injury

1. Remove the pups from their dam and place them in a thermo-regulated environment to avoid hypothermia.
2. Place the pup in the anesthesia induction chamber and begin anesthesia (*see* **Note 1**). If using isoflurane, 2–4 % usually will suffice. The pups will enter different depths of anesthesia. In stage I the pups will become disoriented and may have rapid breathing. In stage II the pups will have irregular respirations with paddling and may flip over on to their backs. In stage III, plane 1, the pups will be under light anesthesia and their breathing will be regular. Pups will still respond to stimulation with movement at this stage. In stage III plane 2, the pups will be similar to plane 1 but will not move when stimulated. An assistant can aid the surgeon and titrate the anesthesia based on the stage and plane of anesthesia.

3. Once the pup is adequately sedated, remove the pup from the chamber and place the pup on the operating board. Place the pup's nose in the nose cone and restart anesthesia. Titrate the isoflurane, if used, from 1 to 2 % to maintain an adequate depth of sedation. To determine the depth of sedation, observe the pup for movement. If the pup is not moving, gently pinch the skin and observe for a reaction. Proceed to the next step if the pup does not react.
4. Secure the pup's limbs with tape to restrain and allow adequate exposure for surgery (Fig. 1a).
5. Sterilize the surgical site by wiping with povidone–iodine wipes three times in a concentric fashion starting at the surgical foci and working outwards. After the povidone–iodine is dry, rinse the surgical site with 70 % isopropyl alcohol. Allow the area to dry to maximize antibacterial effects. Next, drape the pup in a sterile fashion with surgical towels.
6. Make a small incision in the neck (around 1 cm in size) (*see* **Notes 2** and **3**). Under the dissecting microscope, use a blunt dissection technique to expose the neck of the right common carotid artery. Use the curved jeweler's forceps to bluntly dissect the carotid from the surrounding tissue (Fig. 1b).
7. Once the cartotid artery is isolated, electrocauterize it with the cautery pen (*see* **Notes 4** and **5**) (Fig 1c, d).
8. Appose the two ends of the skin wound using the large forceps, and close the small skin wound with Vetbond™.
9. Remove the extra iodine preparation with an alcohol wipe, and place a small amount of triple antibiotic ointment over the wound.
10. Next, mark the pups for proper identification. Our laboratory uses a tattoo system. Prior to the experiment, all pups are assigned an animal number. We then give each pup a tattoo in a unique body location. The tattoo code (for example, a black dot on the right forepaw) then corresponds to the specific animal number. This method enables proper identification of all pups. After tissue removal, the tissue is labeled with the animal number. This number is then used for tissue processing and storage. Using this labeling method, the tissues from an exact pup or experimental group can easily be located.
11. Allow the pups to recover for 30 min in a thermoregulated environment using a heating pad maintained at 37 °C. Place a surgical towel between the pad and the pup along with an Acu-Rite Sensor with an RTD to continuously monitor the temperature.
12. Allow the pups to recover for 2 h with their dam.

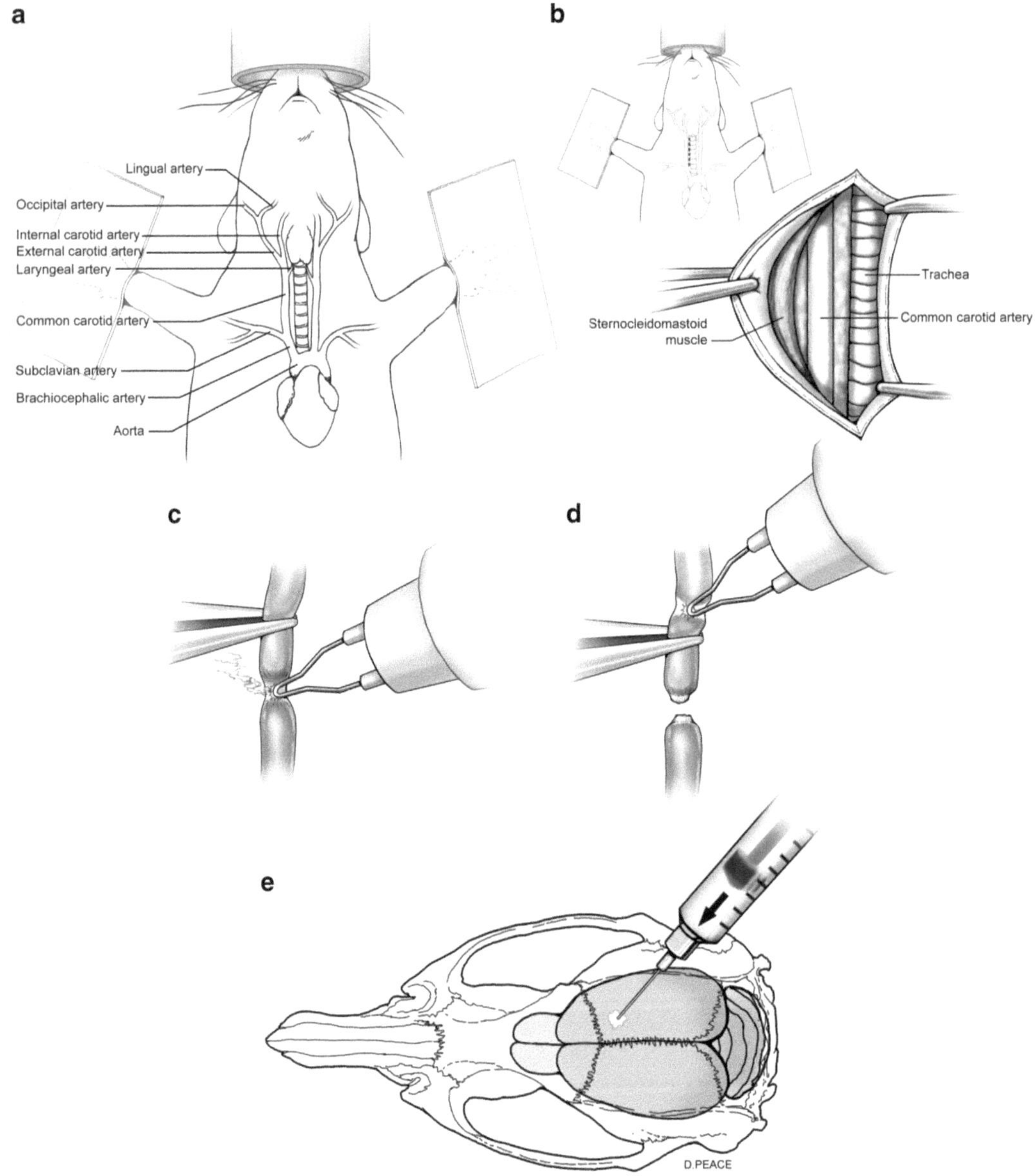

Fig. 1 The ligation of the Rice-Vannucci model and stem cell transplant is shown in illustration form. The common carotid artery is identified (**a**, **b**). The common carotid is carefully grasped with forceps and ligated with a cautery pen (**c**). The top of the artery above the forceps is gently cauterized (**d**). At various times post injury, stem cells are injected into the injured cortex (**e**)

13. Then place the litter of pups in a chamber perfused with a humidified gas mixture of 8 % oxygen and 92 % nitrogen for 60–180 min.
14. Keep the pups' chamber at a constant temperature by using a heated gel pad.

15. After the hypoxic exposure, allow the pups to recover in a temperature-regulated environment using a heating pad (see above).
16. Once fully recovered, return the pups to their dams.

3.2 Preparation of Neurogenic Astrocytes for Transplantation

1. Neurogenic astrocytes are generated from green fluorescence protein (GFP) transgenic neonatal mice (postnatal day 0–6) (The Jackson Laboratory, Bar Harbor, ME, cat# 003116). Animals are anesthetized with hypothermia. Neonatal hypothermia is achieved by putting the pups in a small container, which in turn is placed on ice. To prevent tissue damage, animals are never placed directly on the ice.
2. Decapitate and spray the scalp thoroughly with 70 % EtOH.
3. Following decapitation, remove the brain and dissect the SEZs surrounding the lateral ventricles that contain a high density of NSCs.
4. Mince the tissue chunks with a razor blade, and incubate in 0.25 % trypsin/EDTA. Dissociate the tissue into a single cell suspension. Next, neutralize the trypsin with FBS, pellet the cells and wash them several times in medium. Plate the cells in T75 culture flasks containing growth medium and growth factors.
5. Cells will be passaged 1–3 times before transplantation. Collect confluent monolayers via trypsinization and count with a hemacytometer.
6. Plate a portion of the cell suspension onto polyornithine/laminin-coated glass coverslips to characterize the cells' phenotype prior to transplantation.

3.3 Transplantation of Neurogenic Astrocytes into HIE Models (See Note 6)

1. Prepare the surgical area and the instruments before the pups are removed from the dam to limit the time they are away from the dam.
2. Anesthetize the pups by inhalation of isoflurane.
3. Place the pups in a prechilled homemade clay mold and use illuminating lights for visual guidance to help identify landmarks through the scalp (*see* **Note 7**).
4. Suspend the cells in the base medium or dPBS at a concentration of $5–10 \times 10^4$ cells/μL. Keep the cell suspension on ice during the procedure and gently resuspended with a pipette before loading them into a Hamilton syringe. Make sure air bubbles are not present in the syringe. Push the plunger gently to make sure the cell suspension can be delivered smoothly (*see* **Note 8**).
5. Attach the Hamilton syringe to a micromanipulator and lower the syringe into the cortical areas close to injured sites or into the lateral ventricles of the anesthetized animals (*see* **Note 9**). Slowly inject 1 μL of cell suspension into the targeted areas and leave the needle in place for 3–5 min before removing (Fig. 1e).

4 Notes

1. The mouse pups may require humidification during anesthesia since the dry gas source in combination with the dissecting lights may severely dehydrate the pups.
2. Maintain three separate pairs of surgical instruments clearly marked with different colors of tape. One set of instruments is used for the initial incision, another for the micro-dissections, and a third for the closure. The third set of instruments will retain a small amount of glue from the closure, which can interfere with the microsurgery if the same instruments are used.
3. Begin the procedure with a very small incision (approximately 1 cm) as the incision may become larger with the blunt dissection of the carotid artery. Minimizing the size of the incision will aid in the healing and recovery of the surgical site.
4. A cautery pen greatly decreases surgical time when compared to ligation techniques that use sutures.
5. The cautery pen should be preheated before touching the vessel. If the pen sticks to the vessel during cauterization, keep cauterizing the vessel until the vessel separates completely from the pen. Do not jerk or remove the pen while a fragment of the vessel is still attached. This may cause tearing of the vessel and bleeding.
6. Timing of Transplants: Consider the therapeutic goals and objectives when planning the transplant timing and cell type. After hypoxic–ischemic injury, a variety of pathophysiologic cascades are unleashed. These cascades include inflammation, excitotoxicity, and oxidative stress. Secondary energy failure and cell death also evolve over time [18–20].

   If the transplants are performed in the first *24 h post injury*, the transplanted cells will encounter an environment containing endogenous neuronal necrosis, apoptosis, and secondary energy failure [18, 21, 22]. The therapeutic goal of transplant during the first 24 h after injury may be to interrupt ongoing apoptosis and spare endogenous cells. To achieve these goals, cells such as human umbilical cord blood or an astrocyte line may be chosen.

   At *5–7 days post injury*, the transplanted cells will encounter a brain environment with less apoptosis, a restored energy state, and brain regions cleared of necrosis [18, 21–23]. The goal of transplant 5–7 days after injury may be to repopulate lost neuronal cells and circuitry. This goal may be achieved by transplanting a neuronal cell line. Additionally, a combination of different cells transplanted at different time points may be beneficial. The optimal regenerative medicine approach may require *more than one infusion* of cells or cell types.

The variables discussed above, along with the proper transplant controls (sham-operated and naïve controls) to determine if any cell death of the transplanted cell lines is due to the endogenous milieu should be carefully considered when planning the experimental design.

7. To the best of our knowledge, a stereotaxic apparatus is not commercially available for mouse pups. Instead, we use a homemade clay mold to hold the pups and illuminating lights to help identify landmarks such as sutures and sinuses for visual guidance.
8. An optimal concentration of the cell suspension for the transplantation is important. A concentration too low results in insufficient cell engraftment. Conversely, a concentration too high will cause a dense cellular slurry, which will likely clog the transplant needle. Additionally, donor cells that are crowded and/or forced through the needle results in reduced viability. We have found that a concentration of $5–10 \times 10^4$ works best for the procedure discussed here.
9. Maternal neglect or cannibalism is not uncommon when manipulated pups are returned to their mothers. We have found that dams are more likely to kill and eat their pups if an incision is made. Therefore, cell suspension is injected directly through the scalp and skull to avoid the need for sutures or glues.

References

1. Vannucci RC, Vannucci SJ (1997) A model of perinatal hypoxic–ischemic brain damage. Ann N Y Acad Sci 835:234–249
2. Giffard RG et al (1990) Acidosis reduces NMDA receptor activation, glutamate neurotoxicity, and oxygen-glucose deprivation neuronal injury in cortical cultures. Brain Res 506: 339–342
3. Low JA, Lindsay BG, Derrick EJ (1997) Threshold of metabolic acidosis associated with newborn complications. Am J Obstet Gynecol 177:1391–1394
4. Mulligan JC et al (1980) Neonatal asphyxia. II. Neonatal mortality and long-term sequelae. J Pediatr 96:903–907
5. Azzopardi DV et al (2009) Moderate hypothermia to treat perinatal asphyxial encephalopathy. N Engl J Med 361:1349–1358
6. Gluckman PD et al (2005) Selective head cooling with mild systemic hypothermia after neonatal encephalopathy: multicentre randomised trial. Lancet 365:663–670
7. Shankaran S et al (2005) Whole-body hypothermia for neonates with hypoxic–ischemic encephalopathy. N Engl J Med 353:1574–1584
8. Simbruner G et al (2012) Systemic hypothermia after neonatal encephalopathy: outcomes of neo.nEURO.network RCT. Pediatrics 126:e771–e778
9. Jacobs SE, Tarnow-Mordi WO (2010) Therapeutic hypothermia for newborn infants with hypoxic–ischaemic encephalopathy. J Paediatr Child Health 46:568–576
10. Cilio MR, Ferriero DM (2010) Synergistic neuroprotective therapies with hypothermia. Semin Fetal Neonatal Med 15:293–298
11. Levene MI (2010) Cool treatment for birth asphyxia, but what's next? Arch Dis Child Fetal Neonatal Ed 95:F154–F157
12. Englund U et al (2002) Grafted neural stem cells develop into functional pyramidal neurons and integrate into host cortical circuitry. Proc Natl Acad Sci U S A 99:17089–17094
13. Pimentel-Coelho PM, Mendez-Otero R (2010) Cell therapy for neonatal hypoxic–ischemic encephalopathy. Stem Cells Dev 19:299–310
14. Zheng T et al (2006) Transplantation of multipotent astrocytic stem cells into a rat model of neonatal hypoxic–ischemic encephalopathy. Brain Res 1112:99–105

15. Borlongan CV, Weiss MD (2011) Baby STEPS: a giant leap for cell therapy in neonatal brain injury. Pediatr Res 70:3–9
16. Vannucci RC, Vannucci SJ (2005) Perinatal hypoxic–ischemic brain damage: evolution of an animal model. Dev Neurosci 27:81–86
17. Zheng T et al (2006) Neurogenic astrocytes transplanted into the adult mouse lateral ventricle contribute to olfactory neurogenesis, and reveal a novel intrinsic subependymal neuron. Neuroscience 142:175–185
18. Johnston MV (2001) Excitotoxicity in neonatal hypoxia. Ment Retard Dev Disabil Res Rev 7:229–234
19. Johnston MV et al (2000) Novel treatments after experimental brain injury. Semin Neonatol 5:75–86
20. Ferriero DM (2004) Neonatal brain injury. N Engl J Med 351:1985–1995
21. Northington FJ, Ferriero DM, Graham EM, Traystman RJ, Martin LJ (2001) Early neurodegeneration after hypoxia-ischemia in neonatal rat is necrosis while delayed neuronal death is apoptosis. Neurobiol Dis 8: 207–219
22. Rorke LB (1992) Anatomical features of the developing brain implicated in pathogenesis of hypoxic–ischemic injury. Brain Pathol 2: 211–221
23. Towfighi J, Zec N, Yager J, Housman C, Vannucci RC (1995) Temporal evolution of neuropathologic changes in an immature rat model of cerebral hypoxia: a light microscopic study. Acta Neuropathol 90:375–386

Chapter 14

Isolation and Purification of Self-Renewable Human Neural Stem Cells for Cell Therapy in Experimental Model of Ischemic Stroke

Ricardo L. Azevedo-Pereira and Marcel M. Daadi

Abstract

Human embryonic stem cells (hESCs) are pluripotent with a strong self-renewable ability making them a virtually unlimited source of neural cells for structural repair in neurological disorders. Currently, hESCs are one of the most promising cell sources amenable for commercialization of off-shelf cell therapy products. However, along with this strong proliferative capacity of hESCs comes the tumorigenic potential of these cells after transplantation. Thus, the isolation and purification of a homogeneous, population of neural stem cells (hNSCs) are of paramount importance to avoid tumor formation in the host brain. This chapter describes the isolation, neuralization, and long-term perpetuation of hNSCs derived from hESCs through use of specific mitogenic growth factors and the preparation of hNSCs for transplantation in an experimental model of stroke. Additionally, we describe methods to analyze the stroke and size of grafts using magnetic resonance imaging and Osirix software, and neuroanatomical tracing procedures to study axonal remodeling after stroke and cell transplantation.

Key words Neural stem cells, Stroke, Cell therapy

1 Introduction

Stroke is a serious brain-related pathology worldwide. In the USA, approximately each year 795,000 people suffer from a new or recurrent stroke [1]. Over 800,000 people die in the USA each year from cardiovascular disease and strokes [2]. After stroke, endogenous regeneration through neural stem and progenitor cells is stimulated and newly generated cells actively migrate toward the injured area of the brain [3–9]. However, this cellular turnover does not completely restore function. Current therapeutic approaches, such as the use of thrombolytics, benefit only 1–4 % of patients. Evidence has suggested that cell transplantation therapy might be a viable therapeutic approach for repairing brain damage caused by stroke [10–12]. Normal human-derived somatic stem cells have a limited capacity to proliferate and differentiate into

Brent A. Reynolds and Loic P. Deleyrolle (eds.), *Neural Progenitor Cells: Methods and Protocols*, Methods in Molecular Biology, vol. 1059, DOI 10.1007/978-1-62703-574-3_14,

diverse neural cell types optimal for structural and physiological tissue repair. On the other hand, hESC lines derived from the inner cell mass of preimplantation embryos possess a nearly unlimited self-renewal capacity and the developmental potential to differentiate into virtually any cell type of the organism [13]. This type of stem cells has become an ideal source of cells for regenerative medicine. The process of isolation and purification to obtain a homogeneous cell population of hNSCs is crucial to avoid tumor formation [14, 15].

It has been determined that the differentiation and enrichment processes that direct hESCs towards a neural lineage include: (1) use of defined media supplemented with morphogens or growth factors [15–20], (2) culture under conditions that promote "rosettes", structures morphologically similar to the developing neural tube [21–24], (3) coculture of hESCs with a feeder layer, such as stromal cell lines MS5, PA6, or S17 [25, 26], (4) inhibition of the TGFβ signaling to allow the default neural induction pathway to take place [27–30], or (5) exposure of the hESCs to HepG2 conditioned medium or neural inducing media [31, 32].

This chapter provides a detailed methodology that incorporates the processes as outlined above to derive hNSCs from hESCs, as previously reported [15]. We describe the process to isolate, neutralize, and perpetuate over the long term a homogenous and multipotent hNSC line from hESCs that has demonstrated functional engraftment and no tumorigenicity or overgrowth in experimental model of ischemic stroke [15].

2 Material

Media and solutions must be prepared and all reagents aliquoted. Use ultrapure deionized water to prepare the media and solutions. After prepared, the media containing phenol red must be stored at 4 °C and protected from light.

2.1 Cell Culture Reagents and Solutions

2.1.1 hESC Culture (See Note 1)

1. mTeSR™1 Basal Medium (STEMCELL Technologies; cat. #: 05851).
2. mTeSR™15× Supplement (STEMCELL Technologies; cat. #: 05852).
3. BD Matrigel™ hESC-qualified Matrix (BD, cat. #: 354277).
4. Six-well plates (Corning®; cat. #3335).
5. Collagenase type IV (20 mg/mL): Weight 0.1 g of collagenase type IV (Invitrogen; cat. # 17104-019) and reconstitute in 5 mL DMEM/F12. Filter with 0.2 μm sterile filter system.
6. Dispase (20 mg/mL): Weight 0.1 g of dispase (Invitrogen; cat. #: 17105-041) and reconstitute in 5 mL DMEM/F12. Filter with 0.2 μm sterile filter system (*see* **Note 2**).

2.1.2 hNSC Culture

1. DMEM/F12 + Glutamax (Invitrogen; cat. #: 10565-042).
2. N2 supplement—100× (Invitrogen; cat. #: 17502-048).
3. B27® supplement—50× (Invitrogen; cat. #: 12587-010).
4. Trypsin–EDTA solution (Sigma-Aldrich; cat. #: T4049).
5. Trypsin inhibitor (10 mg/mL) (Sigma-Aldrich; cat. #: T6522): Weight 0.5 g of trypsin inhibitor and dissolve in 40 mL water. Make up to 50 mL, filter with 0.2 μm sterile filter system and aliquot. Store at 4 °C.
6. Human recombinant basic fibroblastic growth factor (bFGF, R & D Systems; cat. #: 233-FB). Stock 10 μg/mL in phosphate-buffered saline (PBS) + 0.1 bovine serum albumin (BSA; Sigma-Aldrich; cat. # A9418) (*see* **Note 3**).
7. Epidermal growth factor (EGF, Millipore; cat. #: 01-101). Stock 20 μg/mL in media.
8. Leukemia inhibitory factor (LIF, Millipore, cat. #: LIF1010). Stock 10 μg/mL.
9. Neural stem cell media (NSC media): Add B27 and N2 supplements to DMEM/F12 + Glutamax. Supplement with EGF (20 ng/ml), bFGF (10 ng/ml), and LIF (10 ng/ml). Store at 4 °C (*see* **Note 4**).
10. Trypan Blue (Sigma-Aldrich, cat. # T8154).
11. Leibovitz's L-15 Medium (Invitrogen; cat. #: 21083-027).

2.1.3 Animal Procedure

1. Cyclosporine A (Sandimmune, Novartis Pharmaceuticals).
2. FORANE (Isoflurane; Baxter; cat. #: 1001936060).
3. Hamilton syringe (5 μL): (Hamilton, cat. #: 87943).
4. Stereotaxic machine (David Kopf Instruments).

2.1.4 Cell Labeling and Magnetic Resonance Imaging

1. Superparamagnetic iron oxide (SPIO) (Feridex IV, Berlex Laboratories).
2. Poly-L-lysine (Sigma-Aldrich; cat. #: P2636).
3. MR imaging: "microSigna 7.0" T-MR scanner (7.0 T/310/AS System; USA: Varion, Inc.).
4. Osirix software (http://www.osirix-viewer.com).

2.1.5 Neuroanatomical Tracing

1. Biotinylated dextran amine 10 % (w/v) (BDA, 10,000 molecular weight (MW); Molecular Probes; cat. #: N7167); Dissolve 1 g of BDA in sterile 10 mL PBS.
2. Peroxidase reaction with diaminobenzidine (Vectastain Elite Kit; Vector Labs; cat. #: PK-6100).
3. Toluidine blue 0.1 % (Sigma-Aldrich; cat. #: T3260): Dissolve 50 mg in 1 N HCl.

4. DPX (distyrene–plasticizer–xylene) mounting medium (Electron Microscopy Sciences; cat. #: 13514).
5. NIH ImageJ Analysis Software (version 1.39 software, http://rsbweb.nih.gov/ij/).

3 Methods

All media and solutions must be warmed to 37 °C before use and all procedures must be performed in sterile environment.

3.1 Generation of hNSC

1. Wash 2× the hESC culture with 1 mL DMEM/F12. Add 0.5 mL collagenase type IV (2 mg/mL) and 0.5 mL dispase (2 mg/mL) solution. Incubate for 2–5 min at 37 °C and 5 % CO_2 (*see* **Note 5**).
2. Gently rinse 2–3× the culture with DMEM/F12.
3. Add 2 mL NSC media and gently scrap the cells with cell scrap or 5 mL pipette.
4. Transfer the cells to conical tube (15 mL). Rinse the well with 3 mL NSC media to remove all remaining cells and transfer to conical tube.
5. Gently pipette up and down 3–5× the NSC media with the cells. Do not dissociate the cells to form single cells. For best results the size of cell aggregate should be in range between 300 and 600 μm.
6. Transfer the aggregated cells to T25 culture flasks (Corning®; 89092-700) and maintain the cells in 7 mL NSC media at 37 °C in a 95 % ar/5 % CO_2 humidified atmosphere.
7. After 3 days in vitro (DIV) expect clusters or spheres (primary spheres; Fig. 1a) of cells to survive and transfer them to a new T75 culture flask with fresh NSC media. Maintain the cells in 7 mL NSC media at 37 °C in a 95 % ar/5 % CO_2 humidified atmosphere and expect the secondary sphere (2° spheres) formation (*see* **Note 6**).
8. Transfer the 2° spheres to a new T25 culture flask with fresh NSC media and maintain the cells for 2 weeks in 7 mL NSC media at 37 °C in a 95 % ar/5 % CO_2 humidified atmosphere (*see* **Note 7**).
9. After 2 weeks, passage the cells with 2 mg/mL collagenase type IV and repeat the procedure every 7 days for additional four passages (total 28 days). Upon completion the cells are fully differentiated into hNSCs with a homogeneous pattern of neural stem/progenitor cell markers (Fig. 1d). Perform phenotype characterization by immunochemistry and RT-PCR to confirm the neural stem cell phenotype (Fig. 1e) [15].

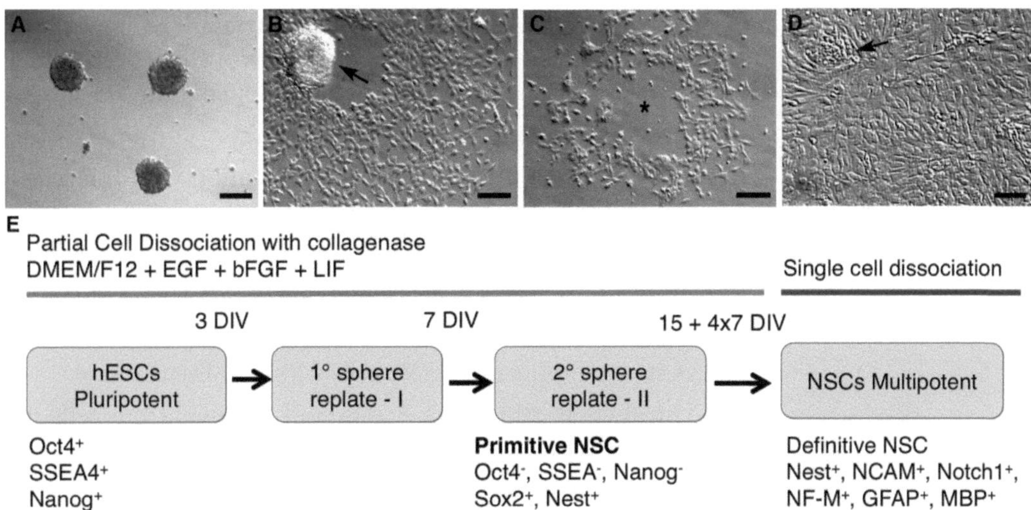

Fig. 1 Isolation and neuralization of hNSCs from the hESCs. (**a**) Primary neurospheres were isolated and replated to eliminate other nonneural cells. The selectively harvested secondary neurospheres (*arrow* in **b**), left behind hollow cores in the surface area (*star* in **c**) where they had been attached. They were perpetuated for an additional five passages (**d**). *Arrow* in (**d**) shows an example of a focus of proliferating cells. (**e**) Schematic representation of the isolation and perpetuation processes of the hNSCs. Neural stem cells were derived from hESCs and propagated using defined media supplemented with EGF, bFGF, and LIF. The developmental progression of the in vitro neural specification and patterning was monitored by the expression of lineage markers as indicated at each stage. Bars: (**a**, **b**, **c**) 200 μm; (**d**) 100 μm. Taken from ref. [15]

10. At this time the hNSCs should get to confluency stage after five DIV and thus are ready for perpetuation on regular basis.
11. To perpetuate the cells, aspirate media and add 1 mL 0.025 % trypsin–EDTA solution for 1 min at 37 °C and 5 % CO_2 (*see* **Note 8**). Next, add an equal volume of 1 mg/mL trypsin inhibitor solution.
12. Transfer the cells to conical tube (15 mL).
13. Gently pipette up and down 3× the solution with the cells to complete cell dissociation.
14. Centrifuge at 300 × *g* for 5 min and discard the supernatant.
15. Gently resuspend the cell pellets in 2 mL of growth Medium.
16. Count the number of viable cells with 4 % Trypan Blue.
17. Calculate the total number of cells/mL.
18. Calculate the total number of cells suspended in the total volume of media.
19. At each passage calculate the fold of increase that is the ratio between the total number of cell and the number of cell plated.
20. Use these data to monitor growth rate of the cells.

3.2 Preparation of Cells for Transplantation

1. Aspirate media and add 1 mL 0.025 % trypsin–EDTA solution for 1 min at 37 °C and 5 % CO_2 (*see* **Note 8**). Next, add an equal volume of 1 mg/mL trypsin inhibitor solution.
2. Transfer the cells to conical tube (15 mL).
3. Gently pipette up and down 3× the solution with the cells to complete cell dissociation.
4. Centrifuge at 300 × g for 5 min and discard the supernatant.
5. Gently resuspend the cell pellets in 2 mL Leibovitz's L-15 Medium.
6. Count the number of viable cells with 4 % Trypan Blue.
7. Centrifuge at 300 × g for 5 min and resuspend the cells to a final concentration of 50,000 cell/μL.

3.3 Cell Transplantation

The surgical procedure for this ischemic model requires the transient occlusion of blood flow of the middle cerebral artery through the introduction of a 4-0 monofilament nylon suture from the external carotid artery to the internal carotid artery, as previously described in detail [33]. According to the experiment, the time of the transplantation may vary.

1. 2 days prior to cell transplantation the rats are immunosuppressed with intraperitoneal (i.p.) injections of cyclosporine A (20 mg/mL). The i.p. immunosuppression continues daily thereafter for 1 week followed by oral cyclosporine at 210 μg/mL in drinking water until perfusion (*see* **Note 9**).
2. Adult male Sprague-Dawley rats weighing 280–320 g are anesthetized during induction with 5% isoflurane that is then reduced to 1–2% during surgery. The anesthesia machine is set to 1.0 L/min O_2 and 1.0 L/min N_2O.
3. 2 μL hNSCs cell suspensions at 50,000 cells/μL are stereotaxically transplanted into four sites localized to the periphery of the infarct in the striatum at the following coordinates: AP: +1.0 mm, ML: +3.2 mm, DV: −5.0; AP: +0.5 mm, ML: +3.0 mm, DV: −5.0; AP: −0.5 mm, ML: +3.0 mm, DV: −5.0; AP: −1.0 mm, ML: +3.5 mm, DV: −5.0 mm with the incisor bar set at 3.4 mm (Fig. 2a).
4. The injection rate is 1 μL/min by Hamilton syringe (5 μL), and the cannula is held in place for an additional 5 min before retraction.

3.4 Magnetic Resonance Image

For tracking hNSCs in vivo by magnetic resonance image (MRI), the cells are labeled with SPIO (*see* **Note 10**) using the poly-L-lysine methods [34] and grafted into the striatum of stroked rats [35].

1. Incubate SPIO and poly-L-lysine in DMEM/12 for 30 min at room temperature while shaking gently.

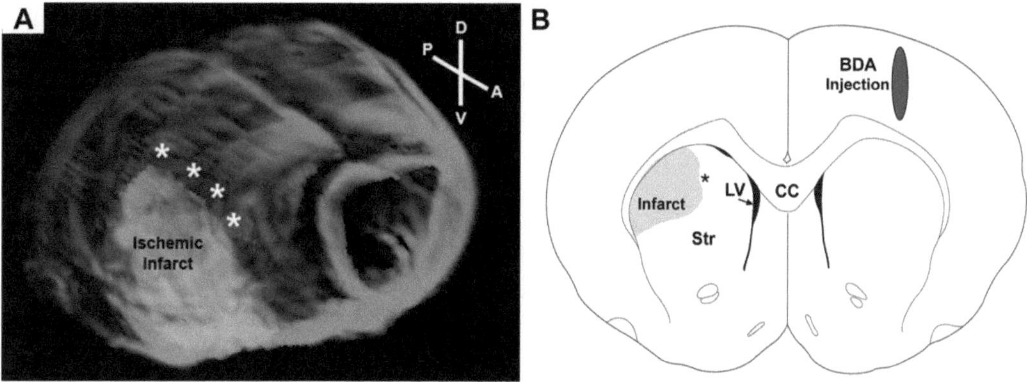

Fig. 2 Sites of hNSC transplantation and BDA injection. (**a**) Three-dimensional surface rendering reconstruction of a rat brain with ischemic injury from high resolution T2-MRI illustrating the ischemia infarct area (hyperintence area). The four *white stars* immediately adjacent to the infarct represent the sites of hNSC transplantation. (**b**) Schematic drawing of a frontal section through the striatum illustrating the focal ischemia-lesioned parenchyma (*shaded area*), graft location (*star*), and the BDA injection site

2. Transfer the SPIO solution to hNSCs and incubate for 72 h.
3. Wash 3× the hNSCs with 10 mL PBS by centrifuge at 300 × *g* for 5 min and resuspend the cell pellets in 1 mL Leibovitz's L-15 Medium.
4. Count the number of cells with 4 % Trypan Blue.
5. Centrifuge at 300 × *g* for 5 min and resuspend the cells according to required cell concentration.
6. MR imaging can be initiated 2 days after transplantation and conducted as long as needed.
7. MR scanner "MicroSigna 7.0" T-MR (7.0T/310/AS System; USA: Varion, Inc., Palo Alto, CA).
8. Anesthetize animals with isoflurane, monitor the respiratory and heart rates and auto-control temperature at 37 °C.
9. Place animals in the resonance instruments: BFG-150/90-S shielded gradient insert (770 mT/m, SR-2500 T/m/s) with a bore size of 9 cm.
10. The imaging protocol consists of imaging in 2D multi-slice followed by a spin-echo sequence.
11. Imaging parameters: TE = 82.5 ms, TR = 4,000 ms, NEX = 10, 5 cm × 5 cm FOV, matrix = 256 × 256, slice thickness = 0.6 mm, gap = 0.

3.5 Infarct and Graft Measurement with Osirix

For transplant volume analysis using the DICOM viewer software Osirix v.3.1 (http://www.osirix-viewer.com) use serial T2-weighted coronal MRI scans with space of 600 μm.

1. The file is first opened in Osirix and the brush tool is chosen to draw the region of interest, i.e., t graft of hNSCs.

2. The hypointense areas of SPIO-labeled cells are measured in the serial scans of the graft.
3. The volume of the transplant is estimated from the total area in the serial sections, space in between and the thickness of the sections.

3.6 BDA Injections and Analysis

To investigate whether the hNSC transplant influenced rewiring of the stroke-damaged side, apply BDA neuroanatomical tracing methodology. 1 week before euthanasia, anesthetize animal and place in the stereotaxic apparatus for stereotaxic BDA injection.

1. After craniotomy, inject 0.5 μL of BDA into the sensorimotor cortex opposite to the stroke lesion site at stereotaxic coordinates: AP: +0.5 m, ML: 2.5 mm, and DV: −1.5 mm (Fig 2b).
2. Close the scalp and return animal to cage.
3. Perfuse animal with 4 % paraformaldehyde 1 week after BDA injection.
4. Perform BDA detection using either immunofluorescence or peroxidase reaction with diaminobenzidine technique.
5. Rinse DAB-stained sections with water and let dry.
6. Coverslip with DPX mounting medium and examine under a light microscope.

For the quantitative analysis of BDA-labeled axon terminals, select three sections approximately 500 μm apart from each experimental case: Use the lateral ventricle, corpus callosum, striatum, and anterior commissure as landmarks for selecting sections [36].

1. Measure the fiber terminals of each scanned field (400 μm^2) using NIH ImageJ Analysis Software.
2. Adjust the background of each image through threshold for only legitimate fibers and puncta.
3. Use ImageJ's "Particle Analyzer" plug-in to identify and measure the number of BDA fiber terminals.
4. Record the number, mean area, mean minimum and maximum pixel intensities for each image (*see* **Note 11**).
5. Using the same sections for axon terminal quantification, determine the total number of BDA+cells at the injection site and use it to normalize the total number of BDA-labeled axon terminals for each experimental case (*see* **Note 12**).
6. Confirm the accuracy of the ImageJ's results by a manual count.
7. Perform the quantitative analysis using two experimenters blinded to all conditions.

4 Notes

1. Prepare mTeSR1 by following manufacturer's instructions: Maintenance of hPSCs in mTeSR™1 and TeSR™2; STEMCELL Technologies. The volume of enzymes and media are applied to one well of six-well plate.
2. Best results are achieved when the collagenase IV and dispase are used immediately after prepared. The enzyme still has activity after frozen (−20 °C), but the incubation time will increase, especially if the enzyme is frozen for a long period. A long period of incubation will decrease number of cells that survive in the experiment.
3. bFGF and all growth factors to be aliquoted under sterile conditions and stored at −20 °C. Avoid filtering media with growth factors due their polarity and protein bounding characteristics.
4. The growth medium is critical in the development of a stable NSC line. Variations may exist from lot-to-lot. This may be due to something as simple as water purity. Thus, it is necessary to set up criteria for testing raw materials and to qualify each of the media components for their consistency in growth promotion.
5. The time of incubation with collagenase type IV and dispase must be adjusted in each passage. The colony edges will appear slight folder back but the colonies remain attached to the plate.
6. Following 5–7 DIV expect the primary spheres to attach to the flask and the fibroblast-like cells to begin to migrate out of the primary spheres (Fig. 1b). At the end of the week the secondary spheres (2° spheres) will lift off and a hollow in the middle of the attached cells can be observed (Fig. 1c).
7. The NSC media is replaced every 3 days and at the end of 2 weeks the cultures reach confluency and are ready for passage.
8. Avoid freeze-thaw 0.025 % Trypsin–EDTA. Make aliquots and store −20 °C.
9. Animal perfusion through the heart: once anesthetized, insert needle directly into protrusion of left ventricle, make a cut in the right atrium and release valve to allow slow, steady flow of around 20 mL/min of 0.9 % saline solution. Allow the solution to flow freely until the blood has been cleared from body and then change to 4 % paraformaldehyde perfusion solution. The extracted brain is then cryopreserved in increasing gradient of sucrose solution (10, 20, and 30%).
10. Feridex IV is no longer being manufactured by Berlex Laboratories. However, an ultrasmall SPIO (USPIO; Ferumoxytol; Feraheme™) is a viable among other alternatives [37].

11. To validate and confirm the computerized counting method of the ImageJ's, conduct a manual count.
12. The total number of BDA-labeled cell bodies at the injection site is critical for normalization and comparative measurement of patterns of connections between cases.

Acknowledgments

Dr Azevedo-Pereira is supported by Coordenação de Aperfeiçoamento de Pessoal de Nível Superior (CAPES, Brazil).

References

1. Roger VL et al (2012) Executive summary: heart disease and stroke statistics – 2012 update: a report from the American Heart Association. Circulation 125:188
2. Minino AM et al (2011) Deaths: final data for 2008. Natl Vital Stat Rep 59:1
3. Arvidsson A et al (2002) Neuronal replacement from endogenous precursors in the adult brain after stroke. Nat Med 8:963
4. Parent JM et al (2002) Rat forebrain neurogenesis and striatal neuron replacement after focal stroke. Ann Neurol 52:802
5. Zhang RL et al (2001) Proliferation and differentiation of progenitor cells in the cortex and the subventricular zone in the adult rat after focal cerebral ischemia. Neuroscience 105:33
6. Zhang R et al (2004) Stroke transiently increases subventricular zone cell division from asymmetric to symmetric and increases neuronal differentiation in the adult rat. J Neurosci 24:5810
7. Jin K et al (2001) Neurogenesis in dentate subgranular zone and rostral subventricular zone after focal cerebral ischemia in the rat. Proc Natl Acad Sci USA 98:4710
8. Jin K et al (2003) Directed migration of neuronal precursors into the ischemic cerebral cortex and striatum. Mol Cell Neurosci 24:171
9. Zhang R et al (2004) Activated neural stem cells contribute to stroke-induced neurogenesis and neuroblast migration toward the infarct boundary in adult rats. J Cereb Blood Flow Metab 24:441
10. STEPS (2009) Stem cell therapies as an emerging paradigm in stroke (STEPS): bridging basic and clinical science for cellular and neurogenic factor therapy in treating stroke. Stroke 40:510
11. Borlongan CV et al (1997) Neural transplantation as an experimental treatment modality for cerebral ischemia. Neurosci Biobehav Rev 21:79
12. Lindvall O, Kokaia Z (2011) Stem cell research in stroke: how far from the clinic? Stroke 42(8):2369–2375
13. Thomson JA et al (1998) Embryonic stem cell lines derived from human blastocysts. Science 282:1145
14. Daadi MM (2011) Novel paths towards neural cellular products for neurological disorders. Regen Med 6:25
15. Daadi MM, Maag AL, Steinberg GK (2008) Adherent self-renewable human embryonic stem cell-derived neural stem cell line: functional engraftment in experimental stroke model. PLoS One 3:e1644
16. Elkabetz Y et al (2008) Human ES cell-derived neural rosettes reveal a functionally distinct early neural stem cell stage. Genes Dev 22:152
17. Koch P et al (2009) A rosette-type, self-renewing human ES cell-derived neural stem cell with potential for in vitro instruction and synaptic integration. Proc Natl Acad Sci USA 106:3225
18. Bain G et al (1995) Embryonic stem cells express neuronal properties in vitro. Dev Biol 168:342
19. Okabe S et al (1996) Development of neuronal precursor cells and functional postmitotic neurons from embryonic stem cells in vitro. Mech Dev 59:89
20. Carpenter MK et al (2001) Enrichment of neurons and neural precursors from human embryonic stem cells. Exp Neurol 172:383
21. Ying QL et al (2003) Conversion of embryonic stem cells into neuroectodermal precursors in adherent monoculture. Nat Biotechnol 21:183
22. Zhang SC et al (2001) In vitro differentiation of transplantable neural precursors from human embryonic stem cells. Nat Biotechnol 19:1129
23. Schulz TC et al (2003) Directed neuronal differentiation of human embryonic stem cells. BMC Neurosci 4:27

24. Cho MS et al (2008) Highly efficient and large-scale generation of functional dopamine neurons from human embryonic stem cells. Proc Natl Acad Sci USA 105:3392
25. Perrier AL et al (2004) Derivation of midbrain dopamine neurons from human embryonic stem cells. Proc Natl Acad Sci USA 101:12543
26. Tabar V et al (2005) Migration and differentiation of neural precursors derived from human embryonic stem cells in the rat brain. Nat Biotechnol 23:601
27. Pera MF et al (2004) Regulation of human embryonic stem cell differentiation by BMP-2 and its antagonist noggin. J Cell Sci 117: 1269
28. Itsykson P et al (2005) Derivation of neural precursors from human embryonic stem cells in the presence of noggin. Mol Cell Neurosci 30:24
29. Reubinoff BE et al (2001) Neural progenitors from human embryonic stem cells. Nat Biotechnol 19:1134
30. Gerrard L, Rodgers L, Cui W (2005) Differentiation of human embryonic stem cells to neural lineages in adherent culture by blocking bone morphogenetic protein signaling. Stem Cells 23:1234
31. Nat R et al (2007) Neurogenic neuroepithelial and radial glial cells generated from six human embryonic stem cell lines in serum-free suspension and adherent cultures. Glia 55:385
32. Shin S et al (2006) Long-term proliferation of human embryonic stem cell-derived neuroepithelial cells using defined adherent culture conditions. Stem Cells 24:125
33. Longa EZ et al (1989) Reversible middle cerebral artery occlusion without craniectomy in rats. Stroke 20:84
34. Frank JA et al (2002) Magnetic intracellular labeling of mammalian cells by combining (FDA-approved) superparamagnetic iron oxide MR contrast agents and commonly used transfection agents. Acad Radiol 9(Suppl 2):S484
35. Daadi MM et al (2009) Molecular and magnetic resonance imaging of human embryonic stem cell-derived neural stem cell grafts in ischemic rat brain. Mol Ther 17:1282
36. Daadi MM et al (2009) Functional engraftment of the medial ganglionic eminence cells in experimental stroke model. Cell Transplant 18:815
37. Castaneda RT et al (2011) Labeling stem cells with ferumoxytol, an FDA-approved iron oxide nanoparticle. J Vis Exp 57:e3482

Chapter 15

Transplantation of Fetal Midbrain Dopamine Progenitors into a Rodent Model of Parkinson's Disease

Lachlan H. Thompson and Clare L. Parish

Abstract

Cell therapy is a promising experimental treatment for Parkinson's disease (PD). It is based on the idea that new dopamine neurons transplanted directly into the forebrain of the patient can structurally and functionally compensate for those lost to the disease in order to restore motor function. While there is a highly active field of research focused on the development of stem cell-based procedures, fetal tissue remains the "gold standard" as a safe and reliable source of dopamine neuron progenitors capable of structural and functional integration with existing motor circuitry following transplantation. This chapter describes the basic procedures for preparation of dopamine progenitor rich cell suspensions of ventral mesencephalon as well as implantation into the unilateral 6-hydroxydopamine model of PD and assessment of functional impact according to drug-induced rotational behavior. The description assumes a basic knowledge of animal handling and stereotaxic surgical procedures in rodents.

Key words Regeneration, Cell therapy, Mesencephalon, Movement disorders, Micro-transplantation

1 Introduction

The inability of the brain to repair itself makes *neural transplantation* an attractive approach for the treatment of neurological conditions. Although the concept has been explored since as early as the late 1800s [1], it gained momentum in the 1970s following studies showing that intra-cerebral grafts of fetal ventral mesencephalon (VM) could restore motor function in a rodent model of Parkinson's disease [2, 3]. These early experiments relied on the use of small VM tissue pieces that were placed in the adult brain by pushing the tissue into resected cortex or injecting it into deeper structures using thin metal cannula. The results showed that the immature dopamine precursors in fetal VM could survive the transplantation procedure while also maintaining their capacity for terminal differentiation and growth into functional mDA neurons.

Brent A. Reynolds and Loic P. Deleyrolle (eds.), *Neural Progenitor Cells: Methods and Protocols*, Methods in Molecular Biology, vol. 1059, DOI 10.1007/978-1-62703-574-3_15, © Springer Science+Business Media New York 2013

Refinement of the procedure has been based largely on improving the survival and integration of dopamine neurons after transplantation while minimizing damage to the host. Important developments include the cell suspension technique that allowed the VM tissue pieces to be dissociated into more or less single cell suspensions prior to transplantation [4, 5] and a micro-transplantation approach using fine glass capillaries to deposit the suspension at precise locations with minimal damage to the host [6]. Cell suspension preparations allow for greater flexibility and reproducibility in transplantation procedures. Fixed volumes at a predetermined cell density can be distributed over multiple graft sites and across multiple animals using the same cell preparation.

Success in animal models of PD lead to the first clinical trials in patients in the 1980s. While the approach has been very successful for some patients [7], the overall variability in the therapeutic outcome [8] highlights the need for optimization of the approach through on-going experimental work in animal models of PD. A deeper understanding of the principles underlying successful restoration of motor function following engraftment of fetal mDA neurons—including survival, growth, and functional connectivity—also forms an important basis for the development of stem cell-based approaches for cell therapy in PD.

In this chapter, we describe the basic procedures for preparation of a VM cell suspension, as illustrated using E12.5 mouse. The dissection procedures are applicable to other species and for other donor ages and, notably, recent studies have suggested younger donor ages may result in better yields of dopamine neurons relative to the conventional choice of E12.5 for mouse and E14.5 for rat [9]. We also describe procedures for lesioning of the nigrostriatal pathway using the monoamine selective neurotoxin 6-hydroxydopamine, implantation of the cell suspension, and assessment of functional impact based on rotational behavior. These procedures are applicable also when using other sources of dopamine neurons for transplantation, such as stem cells.

For recommended further reading on fundamental principles and practical aspects of neural transplantation *see* refs. 10, 11.

2 Materials

2.1 Preparation of Cell Suspensions

2.1.1 Apparatus

1. Dissection microscope (*see* **Note 1**).
2. Dissection instruments including micro-scissors (e.g., spring/vannas, 5–8 mm blades) and Dumont forceps (World Precision Instruments).
3. Hemocytometer and coverslips.
4. Light microscope for basic cell counting.
5. Benchtop centrifuge (preferably refrigerated).

2.1.2 Reagents

1. Cell media including: L15 (Leibovitz) and DMEM$^{-Ca2+/-Mg2+}$, HBSS$^{-Ca2+/-Mg2+}$, PBS$^{-Ca2+/-Mg2+}$ (Invitrogen) or similar.
2. DNase I (Roche). Reconstitute lyophilized protein with dH_2O to a final concentration of 10 mg/ml. Aliquot and store at −20 °C. Dilute 1:200 in medium$^{-Ca2+/-Mg2+}$ for a 0.05 % working solution.
3. Trypsin (Invitrogen). Aliquot 0.25 or 2.5 % stock and store at–20 °C. Dilute ×2.5 or ×25 in medium$^{-Ca2+/-Mg2+}$ for a working solution of 0.1 %.
4. Plastic or glass Petri dish.
5. Plastic Eppendorf tubes.
6. Trypan blue (Invitrogen). For some microscopes with lower light intensity, it may be necessary to dilute the stock Trypan blue 1:10 in PBS in order to distinguish viable (dye-excluding) from dead cells.

2.2 Surgical Procedures and Rotational Behavior

2.2.1 Apparatus

1. Stereotaxic frame with appropriate adaptors for fixed-skull positioning of mice and/or rats (e.g., Stoelting, Kopf, Harvard Instruments).
2. Large probe holder for syringe (Stoelting).
3. Micro-volume syringe with plunger and metal cannula (5–10 μl, e.g., Hamilton or SGE Analytical Science).
4. Surgical microscope, preferably with internal light-source.
5. Suitable handheld drill with appropriate sized drill bit (~1–3 mm).
6. Micropipette puller.

2.2.2 Reagents

1. Borosilicate glass capillaries (Harvard Instruments). Note that the appropriate internal diameter will depend on the outer diameter of the metal cannula on the microsyringe.
2. Heat-shrink tubing, 1.2–1.5 mm (electrical supplies, e.g., RadioParts).
3. 6-hydroxydopamine hydrobromide (6OHDA; Sigma).
4. 0.9% Sterile saline.
5. Ascorbic acid (Sigma).
6. Desipramine (optional; Sigma).
7. Lidocaine (optional; scheduled pharmaceutical).
8. Temgesic (optional; scheduled pharmaceutical).
9. D-Amphetamine (scheduled pharmaceutical).
10. Apomorphine (scheduled pharmaceutical).

3 Methods

All procedures should be performed under aseptic conditions.

3.1 Preparation of Cell Suspensions

1. Remove the uterus immediately from euthanized or terminally anesthetized animals via caesarian section, place in cold L15 medium (*see* **Note 2**) and store on ice.
2. Place the embryos in a petri dish containing ice-cold L15 medium. Under a dissection microscope (*see* **Note 1**), remove each embryo using suitably fine dissection instruments. This is best achieved through a single incision parallel to the placental mass (*see* Fig. 1a, b) while applying gentle pressure with the forceps to both sides of the sac—the positive pressure in the amniotic sac will usually expel the embryo. Cut the umbilical cord (Fig. 1c) and repeat for the next embryo.
3. The ventral mesencephalon can be obtained from each embryo in five steps using forceps to stabilize the embryo and scissors to isolate the VM as a single piece of tissue (Fig. 1):
   (a) The anterior end of the mesencephalic tube is cut at approximately the di-mesencephalic boundary—parallel and directly adjacent to where the narrow tube expands into the larger forebrain regions containing the developing cortex and thalamus (Fig. 1e).
   (b) The caudal end of the mesencephalic tube is cut at approximately the level of the mid-hindbrain boundary (Fig. 1e; *see* **Note 3**).
   (c) The mesencephalic tube can now be isolated by sliding one blade of the scissors between the meningeal layers and the VM—perpendicular to the cuts performed in **step 3** (a, b)—in order to cut the connective tissue overlaying the midbrain (Fig. 1f, g).
   (d) The ventral midline is exposed by "rolling" the VM tube 90° away from the user (Fig. 1h). The dorsal midline will remain attached to the meningeal tissue, face-down on the petri dish. Cut parallel to the midline approximately 1/3 of the distance between the ventral and dorsal midline on one side of the VM tube (Fig. 1h).
   (e) Perform a second cut at the same position on the other side of the tube to free the final VM piece used for transplantation (Fig. 1h; *see* **Note 4**). It may be necessary to trim excess tissue and it is important to peel away any remaining meningeal tissue (*see* **Note 5**).
4. Remove the media from the eppendorf containing the pooled VM pieces and replace with medium that does not contain Ca^{2+} or Mg^{2+} (e.g., DMEM/PBS/HBSS$^{-Ca2+/-Mg2+}$) and is sup-

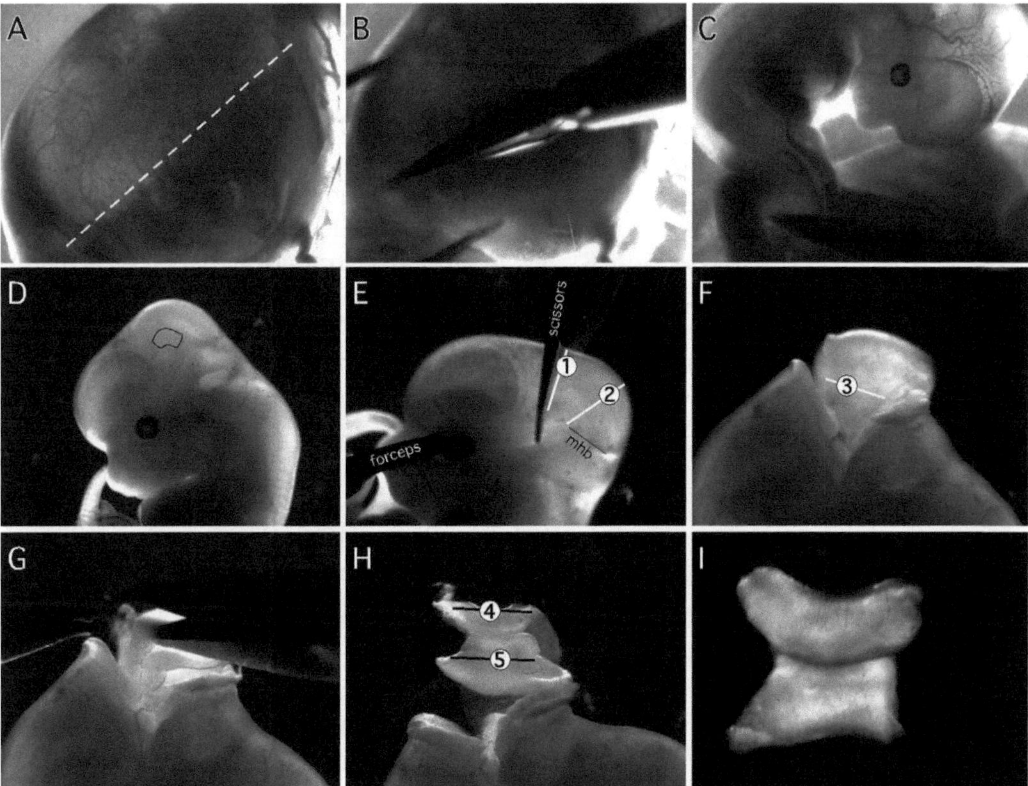

Fig. 1 Dissection of ventral mesencephalon from E12.5 mouse. (**a**) The placental mass can be seen through the intact embryonic sac using a dissection microscope with a backlit stage. The *dashed line* illustrates a convenient location to make an incision (**b**) using micro-scissors and forceps. (**c**) The umbilical cord is cut once the embryo is expelled from the embryonic sac. (**d**) The *grayed-out* region highlights the approximate area of the ventral part of the neural tube as the target for dissection. (**e**) The *white lines* indicate the anterior (*1*) and posterior (*2*) dissection planes for isolating the midbrain from the embryo. The mid-hindbrain border (mhb) is indicated in *black*. (**f**) The connective meningeal tissue can be separated from the underlying VM through an incision perpendicular to the first two cuts (*3*) by sliding a blade in between these tissue layers (**g**). Rolling the now isolated midbrain tube 90° away from the user will expose the ventral floor plate (**h**). The ventral part of the midbrain can now be isolated by two parallel cuts as indicated by the *black lines* (*4, 5*). (**i**) The final dissected VM piece has a characteristic "butterfly" shape

plemented with 0.05 % DNase and 0.1 % trypsin. Incubate at 37 °C for 20 min (*see* **Note 6**).

5. Rinse the VM pieces in medium$^{-Ca2+/-Mg2+}$ three times before adding a known volume of medium$^{-Ca2+/-Mg2+}$ with 0.05 % DNase for tissue dissociation (*see* **Note** 7).
6. Mechanically dissociate the VM tissue pieces by gently trituration with a 1 ml pipette tip followed by a 200 μl tip. Continue as necessary to achieve a cloudy suspension.
7. To estimate the total viable cell number, dilute a small (e.g., 1–5 μl) sample of the suspension 1:10 in Trypan blue and

count viable (dye-excluding) cells using a hemocytometer. Total viable cells = cells counted × dilution factors (e.g., trypan dilution and fractional area of hemocytometer counted) × volume of suspension. Cell viability should be >90%.

8. Centrifuge at 500 × *g* for 5 min at 4 °C. Remove the supernatant and use a pipette to resuspend the pellet in medium$^{-Ca2+/-Mg2+}$ with 0.05 % DNase at the desired cell density (*see* **Note 8**). Be careful not to introduce air bubbles when resuspending. Store on ice throughout transplantation procedure.

3.2 Unilateral Lesioning of the Nigrostriatal Dopamine System in Rodents

1. Prepare the 6-hydroxydopamine (6OHDA) solution by dissolving 6OHDA in sterile saline (0.9 %) containing 0.02 mg/ml L-ascorbic acid (Sigma) to achieve the desired concentration (*see* **Note 9**). Do not use PBS, this will oxidize the toxin (*see* **Note 9**).
2. Rodents should be deeply anesthetized with the head placed in a fixed-skull position using a stereotaxic frame.
3. Make an incision in the scalp along the midline and drill a small burr hole in the skull at the desired coordinates (*see* **Note 10**) to reveal the dural surface.
4. Inject the 6OHDA solution using a micro-volume syringe, preferably fitted with a pulled glass cannula (*see* **Note 11**). Leave the cannula in place for at least 2 min after injection to allow for diffusion of the toxin. If necessary, the central noradrenergic system can be protected by intraperitoneal injection of desipramine (20 mg/kg; Sigma) 30 min before injection of 6OHDA (*see* **Note 12**).
5. Close the scalp (e.g., using suture or Michel clips). A topical anesthetic (e.g., lidocaine) and/or a central analgesic (e.g., Temgesic 0.3 mg/kg) may be administered. This is particularly relevant for short recovery times, e.g., when using inhaled anesthesia.

   The functional impact of unilateral lesioning of the nigrostriatal pathway can be readily determined by measuring turning behavior in response to D-amphetamine or apomorphine.
6. For D-amphetamine-induced rotation, dissolve D-amphetamine in 0.9 % sterile saline and administer (i.p.) 2.5 mg/kg for rats or 5 mg/kg for mice. For apomorphine-induced rotation, dissolve apomorphine in 0.9 % saline with 0.02 mg/ml L-ascorbic acid administer (s.c.) 0.025 mg/kg for rats or 0.1 mg/kg for mice (*see* **Note 13**).
7. Allow the animals to habituate for 10 min after drug injection. Record the number of full (360°) body rotations in each direction over a defined time period—typically 40–60 min for apomorphine and 60–90 min for amphetamine (*see* **Note 14**).

8. Record the rotation score for each animal as net full body rotations per minute in one direction (*see* **Note 15**).

3.3 Implantation of Cell Suspension

1. Prepare deeply anesthetized animals in a sterotaxic frame and drill a burr hole through the skull at the desired coordinates as per **steps 2** and **3** of Subheading 3.2 (*see* **Note 16**).
2. Resuspend the cells through gentle titruation using a pipette to achieve a uniform suspension. It is important to repeat this step prior to each implantation as an effort to deliver a consistent number of cells across all animals.
3. Prefill the microsyringe and glass cannula with vehicle (e.g., 0.9 % saline or cell medium) and draw the required volume of cells into the glass cannula through the tip (unlike when using aqueous solutions, there is no need to draw an air-bubble first).
4. Slowly pressure-inject the cells at the desired location. It is important to monitor cell-flow through the glass cannula throughout the injection to ensure that all the suspension has been injected (this may not be the case if there is a blockage or a leak in the cannula setup) and, importantly, that there is no backflow of the injected contents during the injection (*see* **Note 17**). Keep the cannula in place for at least 2 min before slowly withdrawing.
5. Close the scalp and administer local anesthetic and central analgesia as per **step 5** of Subheading 3.2 above.

4 Notes

1. The dissection microscope should allow sufficient working distance between the stage and objective for manipulation of the embryos. It is helpful to use a backlit (beneath the stage) system with an adjustable mirror to give a range of contrast options. The addition of a halogen lamp and optical filters for fluorescence is particularly useful when working with transgenic mice expressing fluorescent proteins.
2. We routinely use L15 medium (Invitrogen), although any physiological salt solution may be acceptable—e.g., HBSS, DMEM, PBS, NaCl. Generally, quick dissections (i.e., <20 min) may be done under minimalist conditions (e.g., PBS, NaCl), while longer dissections are more likely to benefit from media containing nutritional supplements including salts and amino acids to support cell viability.
3. Grafts of ventral mesencephalon will invariably contain a fraction of serotonergic neurons originating from the developing raphe nucleus, along the ventral midline immediately caudal to the mid-hindbrain border. The position of the caudal dissection

of the mesencephalic tube will thus impact on the amount of serotonergic progenitors included in the dissected tissue piece. This is relevant to mention in light of recent studies suggesting that high numbers of serotonergic neurons in striatal VM grafts may contribute unwanted side effects such as dyskinesias [12, 13]. Positioning the caudal dissection forward of the mid-hindbrain border will limit the contribution of serotonergic progenitors in the final cell preparation.

4. Move the dissected VM pieces away from the stage light and into a darker area towards the side of the petri dish. When dissecting large numbers of embryos, periodically move the dissected VM pieces to an eppendorf containing ice-cold cell culture medium (L15, HBSS, DMEM, or equivalent).
5. The meningeal tissue can be difficult to remove at earlier embryonic ages (i.e., <E13.5 for rat and <E11.5 for mouse). At these earlier ages, it is useful to incubate the dissected VM tissue piece(s) with collagenase/dispase (700 mg/ml; Roche) for 10 min at 37 °C before attempting to remove the meninges.
6. Pre-warm the incubation medium to 37 °C. Various cell culture media not containing Mg^{2+} or Ca^{2+} are available (e.g., $HBSS^{-Ca2+/-Mg2+}$, $DMEM^{-Ca2+/-Mg2+}$, $PBS^{-Ca2+/-Mg2+}$) and may be acceptable. We routinely use $DMEM^{-Ca2+/-Mg2+}$.
7. Add at least 10–20 μl of medium per VM piece. Insufficient volumes will result in clumping of the cells and present difficulties when preparing the final suspension for grafting.
8. We typically resuspend at 1×10^5 cells/μl in a PCR tube or small eppendorf. Using smaller tubes will reduce the risk of breaking the glass cannula when loading the cells (**step 3** of Subheading 3.3). We have found densities $>1.5 \times 10^5$ cells/μl to occasionally be problematic during transplantation due to cell clumping and blockage of the glass injection cannula. For viability of the cell suspension, we have also found it best to work with larger volumes (>10) at the expense of cell density—i.e., in order to transplant 1×10^5 cells, consider preparing 10 μl at 0.5×10^5 cells/μl, and injecting 2 μl rather than 5 μl at 1×10^5 cells/μl and injecting 1 μl.
9. 6OHDA is light-sensitive and prone to oxidation. To minimize exposure to light and moisture, it is useful to aliquot stock 6OHDA into eppendorf tubes and note the amount of 6OHDA on the tube lid (e.g., 1–3 mg). Store aliquots in the dark at −20 °C. Change working solutions of 6OHDA every 2 h (ongoing oxidation of the toxin in solution is evident as a progressively deeper orange/brown color). The 6OHDA is generally supplied as a salt derivative (e.g., only 82 % free-base when supplied as 6OHDA hydrobromide, Sigma-Aldrich) and this should be accounted for when calculating the final dilution.

The biological efficacy of 6OHDA stocks can show significant batch variation. When working with a new stock, it is recommended to test efficacy in a small group of animals (based on rotational behavior and/or immunohistochemistry for tyrosine hydroxylase) before undertaking large-scale experiments.

10. The dosage and placement of 6OHDA will depend on the aims of the experiment. Optimal stereotaxic coordinates may vary with the strain, sex, and age of the animal and should be determined by the user. For adult *rats*, complete lesioning of the nigrostriatal system (substantia nigral (SN) and ventral tegmental (VTA) dopamine neurons) can be achieved by injection of 14 μg of 6OHDA (4 μl of a 3.5 μg/μl solution) into the medial forebrain bundle of adult rats at the following stereotaxic coordinates: 4.4 mm caudal to bregma on the anterior–posterior axis (AP −4.4); 1.2 mm lateral to the midline (ML ±1.2); 7.8 mm below the dural surface (DV −7.8). Partial lesions (largely sparing the VTA neurons) can be achieved by multiple 6OHDA injections (2 μl of a 3.5 μg/μl solution per site) into striatal target areas of the SN neurons (AP +1/−0.1/−1.2; ML ±3/3.7/4.5; DV −5). For adult *mice*, lesioning of nigrostriatal dopamine pathway can be achieved through injection of 6OHDA directly into the midbrain (AP −3.0, ML ±1.2, DV −4.5) and the severity of the lesioning can be modulated by the dose of 6OHDA—e.g., 1.5 μl of a 1.5 μg/μl solution for partial lesioning and 2 μl of a 3 μg/μl solution for more complete lesioning. The results can vary markedly in mice and optimal dosage should be determined by the user. Mortality in mice (up to 50 %) can become an issue at larger doses for complete lesioning of the midbrain dopamine neurons.

11. The use of fine glass cannulas is recommended for intracranial injection procedures in rodents as a means to reduce physical damage and reflux of the injected material. Standard borosilicate glass capillaries (Harvard Instruments) can be pulled into fine glass cannula using a micropipette puller. The capillary can then be fitted to the end of a micro-volume syringe using heat-shrink tubing (Fig. 2). Before injecting the toxin, fill the micro-syringe and attached glass cannula with vehicle (sterile 0.9 % saline) using a standard syringe to inject the microsyringe from the top before replacing the plunger. Use the plunger to draw up an air-bubble of 0.5–1 μl before drawing up the required amount of toxin from a small eppendorf. This will prevent dilution of the toxin by diffusion through the vehicle.

12. 6-hydroxydopamine selectively destroys catecholaminergic neurons, including both dopamine and noradrenergic neurons. Although the ablation of noradrenergic neurons is largely overlooked in 6OHDA procedures targeting the nigrostriatal

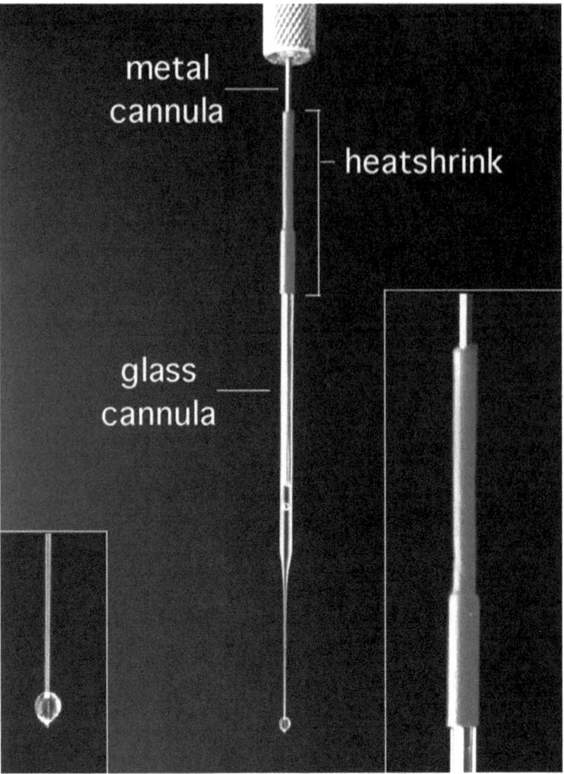

Fig. 2 Glass injection cannula fixed to a micro-volume syringe. The glass cannula can be fitted over the metal cannula of a micro-syringe using a sleeve of heat shrink (*red*). The sleeve is heated (e.g., using a handheld blowtorch) to form a water-tight seal between the glass and metal cannula. It is important to flush the syringe thoroughly with vehicle to ensure no leakage from either end of the heat-shrink seal before proceeding. The *enlarged sections* show the seal as well as a drop of saline at the tip of the glass cannula in greater detail. Note that it is also possible to use appropriately sized polyethylene tubing to join the glass and metal cannulas. In that case, the polyethylene is first fitted to the metal cannula and the glass is passed over the polyethylene tubing. It is normal to shear the polyethylene tubing to form a water-tight seal

dopamine system, it may be important to protect the integrity of the noradrenergic system in certain cases depending on the specific aims of the experiment.

13. For apomorphine-induced rotation, all animals should be treated at least twice to allow for sensitization of the rotational response before recording rotational data on the third administration. It is recommended to allow a 48 h rest period between treatments. This is only required at the beginning of the experiment and subsequent rotational measurements in long-term experiments will only require a single apomorohine treatment.

14. Rotational behavior is most conveniently assessed using automated "rotometer" systems (e.g., Accuscan Instruments) whereby a harness placed around the animal is connected to a transducer that records the number and direction of rotations over time. Alternatively, rotations can be scored manually under suitable single-housing conditions, e.g., home cage or a large cylinder.
15. Unilateral lesioning of the nigrostriatal system will result in turning behavior *towards* the side of the lesion following amphetamine and *away* from the lesioned hemisphere following apomorphine. It is advisable to set a threshold response level of ≥6 turns per minute as an inclusion criterion when selecting animals for further experimentation. Lack of stability of the rotation response over time may become an issue when using animals with lower rotation scores. When determining animal numbers for experimental grouping, it should also be noted that there is invariably a population of "nonresponders" (up to 50 % depending on and species and strain). This aspect can be particularly problematic when using mice. A number of previous studies have provided detailed descriptions of the impact of 6OHDA lesioning on the nigrostriatal dopamine system and motor behavior in rats [14, 15] and mice [16, 17].
16. Optimal stereotaxic coordinates should be determined by the user. As a guide, intra-striatal placement can be achieved at the following coordinates: adult rats (AP +0.5, ML ±3, DV -4.6); adult mice (AP +1.0, ML ±2.3, DV -3.2). Grafting into the *substantia nigra* can be achieved using the following coordinates: adult rats (AP -5.3, ML ±2.2, DV -7.2); adult mice (AP -3.2, ML ±1.4, DV -4.2).
17. Backflow of single cell suspensions is an important but often overlooked issue in neural transplantation procedures. One should carefully monitor the point at which the cannula enters the brain to ensure there is no significant backflow of the suspension up the cannula and onto the surface of the brain. Ensuring the tip of the cannula is "flat" (rather than jagged and broken) may help to prevent backflow. This can be achieved by scoring and snapping the end of the cannula. If backflow presents as a consistent problem during surgery, it is recommended to replace the cannula.

References

1. Thompson WG (1890) Successful brain grafting. Science 16:78
2. Perlow MJ et al (1979) Brain grafts reduce motor abnormalities produced by destruction of nigrostriatal dopamine system. Science 204:643
3. Bjorklund A, Stenevi U (1979) Reconstruction of the nigrostriatal dopamine pathway by intracerebral nigral transplants. Brain Res 177:555
4. Bjorklund A, Schmidt RH, Stenevi U (1980) Functional reinnervation of the neostriatum in the adult rat by use of intraparenchymal

grafting of dissociated cell suspensions from the substantia nigra. Cell Tissue Res 212:39
5. Bjorklund A et al (1983) Intracerebral grafting of neuronal cell suspensions. I. Introduction and general methods of preparation. Acta Physiol Scand Suppl 522:1
6. Nikkhah G et al (1994) Improved graft survival and striatal reinnervation by microtransplantation of fetal nigral cell suspensions in the rat Parkinson model. Brain Res 633:133
7. Lindvall O, Hagell P (2000) Clinical observations after neural transplantation in Parkinson's disease. Prog Brain Res 127:299
8. Winkler C, Kirik D, Bjorklund A (2005) Cell transplantation in Parkinson's disease: how can we make it work? Trends Neurosci 28:86
9. Torres EM et al (2007) Improved survival of young donor age dopamine grafts in a rat model of Parkinson's disease. Neuroscience 146:1606
10. Bjorklund A, Dunnett SB (1994) Functional neural transplantation II. Novel cell therapies for cns disorders. Progress in brain research, vol 127. Elsevier, New York
11. Dunnett SB, Boulton AA, Baker GB (2000) Neural transplantation methods. In: Wolfgang W (ed) Neuromethods, vol 36. Humana, Totowa, p 576
12. Carlsson T et al (2007) Serotonin neuron transplants exacerbate L-DOPA-induced dyskinesias in a rat model of Parkinson's disease. J Neurosci 27:8011
13. Bjorklund A et al (1983) Intracerebral grafting of neuronal cell suspensions. VII. Recovery of choline acetyltransferase activity and acetylcholine synthesis in the denervated hippocampus reinnervated by septal suspension implants. Acta Physiol Scand Suppl 522:59
14. Kirik D, Rosenblad C, Bjorklund A (1998) Characterization of behavioral and neurodegenerative changes following partial lesions of the nigrostriatal dopamine system induced by intrastriatal 6-hydroxydopamine in the rat. Exp Neurol 152:259
15. Lee CS, Sauer H, Bjorklund A (1996) Dopaminergic neuronal degeneration and motor impairments following axon terminal lesion by instrastriatal 6-hydroxydopamine in the rat. Neuroscience 72:641
16. Iancu R et al (2005) Behavioral characterization of a unilateral 6-OHDA-lesion model of Parkinson's disease in mice. Behav Brain Res 162:1
17. Lundblad M et al (2004) A model of L-DOPA-induced dyskinesia in 6-hydroxydopamine lesioned mice: relation to motor and cellular parameters of nigrostriatal function. Neurobiol Dis 16:110

Part V

In Vivo Identification of Neural Precursor Cells

Chapter 16

Distribution of Neural Precursor Cells in the Adult Mouse Brain

Daniel G. Blackmore and Rodney L. Rietze

Abstract

Since its inception in 1992 [Reynolds and Weiss, Science 255:1707–10, 1992], the neurosphere assay (NSA) has proven an exceptionally useful tool in detecting neural stem cells (NSCs) in both the developing and adult mammalian brain. To date, over 1,300 manuscripts have been published employing the assay, attesting to the robustness of the assay, and its ease of use. However, a brief survey of the literature demonstrates that the number of primary neurospheres generated from essentially the same anatomical region (i.e., the periventricular region of the rostral lateral ventricle) ranges between 150 and 936 [Gritti et al., J Neurosci 22:437–445, 2002; Tropepe et al., J Neurosci 17:7850–59, 1997; Doetsch et al., Cell 97:703–16, 1999; Enwere et al., J Neurosci 24:8354–65, 2004]. Indeed, in our hands we typically generate approximately 1,800 primary spheres when harvesting tissue from the same region.

Key words Neural stem cell, Progenitor cell, Precursor, Neurosphere, Vibratome section, Subventricular zone, Neuraxis

1 Introduction

Following our discovery that a 1:1 relationship does not exist between NSCs and neurospheres [6] and in association with the development and validation of the neural-colony forming cell assay (N-CFCA)—which can discriminate between NSCs and progenitors [7]—we performed a comparative analysis of the three most widely used techniques (currently employed) to detect NSCs and progenitor cells. Given a gross dissection (as typically performed) was not adequate for such an analysis, coupled with the vast differences in number of primary spheres generated between labs using essentially the same technique, we chose to serially section adult mouse brains into 400 μm segments along the ventricular neuraxis utilizing definitive and easily identifiable landmarks to ensure that the same regions of the brain were consistently obtained [8].

Using this improved methodology, we found a large variation in the frequency and distribution of NSCs and progenitor cells

Brent A. Reynolds and Loic P. Deleyrolle (eds.), *Neural Progenitor Cells: Methods and Protocols*, Methods in Molecular Biology, vol. 1059, DOI 10.1007/978-1-62703-574-3_16, © Springer Science+Business Media New York 2013

(both of which have the ability to form neurospheres) along the neuraxis, with the greatest variability in the rostral portion of the lateral ventricles, thereby explaining the large variation in neurosphere frequency previously reported. Furthermore, we found specific coordinates, corresponding to the cephalic flexures during development (boundaries of the fore, mid, and hindbrain) where NSCs were absent, suggesting that factors which control regional specialization also influence the generation of distinct populations of NSCs. In light of prior work describing the disparate responsiveness of NSCs to various growth factors along the rostral–caudal axis, the latter finding is also not unexpected. Finally, we found that our harvesting of smaller pieces of tissue required a less harsh dissociation procedure. This resulted in the generation of significantly greater numbers of neurospheres from the same anatomical region previously reported, providing a more complete picture of the frequency and possibly the diversity of neural precursor cells in the brain.

Here we provide a detailed description of all aspects of the procedure, from the removal of the brain from the adult mouse (C57/Bl6J mice), to the transfer of cells to the appropriate functional assay. We have assumed that the reader possesses a fundamental understanding of neuroanatomy, but for further reference, we strongly suggest employing a mouse brain atlas or similar online resource [9, 10].

2 Materials

2.1 Reagents

1. Adult mice (6–8 weeks, C57BL/6J).
2. 10× Phosphate-buffered saline (PBS).
3. 70% (vol/vol) ethanol.
4. Basic fibroblast growth factor (bFGF, 10 ng/ml).
5. Bovine serum albumin (Sigma—A3311).
6. Bromodeoxyuridine (BrdU, Sigma—B5002-5G).
7. Calcium chloride (Sigma—C1016-500G).
8. Complete medium (CM): mouse Neurocult™ NSC Basal medium + Proliferation supplement (Stem Cell Technologies).
9. Cutting solution 10× Stock.

   In a 1 l volumetric flask add the reagents below making sure that the $CaCl_2$ is added last.

   Start by placing 69.5 g NaCl into flask then add 750 ml of MilliQ water then add the following components.

   1 M KCl: 25 ml.

   1 M NaH_2PO_4: 12 ml.

$MgCl_2$: 13 ml.

$CaCl_2$: 25 ml.

Make sure the $CaCl_2$ is added last to the solution. Mix solution until dissolved. Fill flask to 1 l and then transfer to 1 l bottle and store at 4 °C until needed. The 10× solution can be stored for approximately 1 month.

10. *Working Cutting Solution.* On the day before use if 500 ml is needed (this is enough for three brains) measure out 50 ml of the 10× stock and add to a 500 ml bottle. Add 450 ml of milliQ water. Then add 1.1 g Sodium bicarbonate ($NaHCO_3$) and 1 g D-glucose. Mix well. Place the solution at 4 °C for the following day. Filter sterilize cutting solution using 500 ml Nalgene/Nunc Filtration Unit and place into −80 °C for approx. 45 min until an ice slush is achieved. Use a metal spatula to mix solution well making sure ice is removed from edge of flask and included in the ice solution.
11. Epidermal growth factor (EGF)—receptor grade (BD Biosciences, Australia).
12. Glucose (Sigma—G7021-1KG).
13. HCl (Ajax Finechem—AF702102).
14. Heparin.
15. HEPES.
16. HEPES-buffered minimum essential medium (HEM).
17. Magnesium chloride (Sigma—M8266-100G).
18. MilliQ water.
19. Normal Goat Serum.
20. Neural-Colony Forming Cell Assay kit (Stem Cell Technologies—05740).
21. Neurocult™ NSC Basal Medium (Stem Cell Technologies—05702).
22. PAP pen (Sapphire Bioscience, Sydney, Australia).
23. Paraformaldehyde (Sigma—P6148-500G).
24. Penicillin/streptomycin—100 U/ml (Gibco/Invitrogen).
25. Petri dishes: 100 × 15 mm (BD Falcon—351029).
26. Potassium chloride (Sigma—P3911-500G).
27. Sodium azide (Sigma—S2002-25G).
28. Sodium bicarbonate (Sigma—S5761-500G).
29. Sodium chloride (Sigma—S9888).
30. Sodium dihydroxide phosphate (NaH_2PO_4) (Ajax Finechem—A471-500G).
31. Sodium pentobarbitone (Lethabarb Virbac—1PO643-1).

32. Sucrose (Sigma—S9378).
33. SuperFrost Plus slides (Menzel-Glaser—J1800 AMNZ).
34. Trypan Blue (Sigma—T8154).
35. Trypsin–EDTA—0.1 %.
36. Trypsin inhibitor—0.014 % w/v (type I-S from soybean: Sigma-Aldrich).

2.2 Equipment

1. 70 μm sieve (Falcon/BD Biosciences).
2. Cell culture dish with 2 mm Grid (Nunc—1174926).
3. Centrifuge.
4. Eppendorf tubes: 1.5 ml.
5. Falcon tubes: 15 ml (BD Falcon—352096).
6. Falcon tubes: 50 ml (BD Falcon—352070).
7. Glass bead sterilizer (Fine Science Tools).
8. Gloves (Pharmatex—ARTG 127794).
9. Humidified tissue culture incubator (37 °C, 5 % CO_2).
10. Kimwipes—42.5 × 37.5 cm (Kimberly-clark—4108).
11. Kimwipes—21 × 12 cm (Kimberly-clark—4103).
12. Laminar flow hood.
13. Leica VT 1,000 s vibratome.
14. Light source: fiber optic.
15. Magnetic mixer.
16. Magnetic stirring rod.
17. Media bottles (500 ml, 1 l, 2 l).
18. MICROM cryostat.
19. Microdissection instruments: Size 3 scalpel holders (×3), 11.5 cm straight dissection scissors, fine straight forceps, and curved fine forceps (Fine Science Tools).
20. Nunc Filter Unit (Nalgene, cat. no. NAL5660020).
21. Paint Brush-fine.
22. Biosafety level two tissue culture hood.
23. Pippette tips: Filter-tipped (20 μl, 200 μl, 1 ml).
24. Razor Blades: Double edged.
25. Scalpel blades: size 10.
26. Serological Pipet: 5 ml (Falcon—35 7543).
27. Serological Pipet: 10 ml (Falcon—35 7551).
28. Serological Pipet: 25 ml (Falcon—35 7525).
29. Syringe Needle: $25^{5/8}$ Gauge (BD Precision Glide—301 805).
30. Stereo dissection microscope.

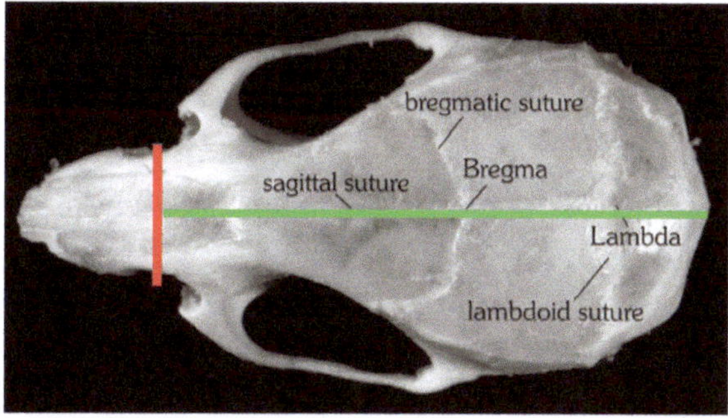

Fig. 1 Dorsal surface of the mouse skull showing the location where the skull should cut (*red line*, first cut; *green line*, third cut) to enable the brain to be removed. Figure reproduced from "The mouse brain in stereotaxic coordinates" [10] with permission from Academic press

31. Storage Media Bottle 150 ml (Corning—431175).
32. Super Glue Gel (UHU GmbH and Co.).
33. Syringe: 1 ml (Terumo—SS+01T).
34. Tissue culture flasks: T25 (Nunc, cat. no. 0602).
35. Volumetric flask: 2 l.
36. Water bath at 37 °C.

3 Methods

3.1 Tissue Isolation

1. Sacrifice mouse via ethically approved method that minimizes any mechanical damage to the skull (i.e., cervical dislocation).
2. Place mouse on absorbent tissues (dorsal side up) and sterilize the head and body by spraying with 70 % (vol/vol) ethanol.
3. Place fingers on either side of the skull, then use a scalpel to cut the skin longitudinally along rostral–caudal axis of the skull from about the olfactory bulb to the level of the cerebellum at the base of the skull.
4. To begin to remove the brain from the skull, dissect open the skull by first using fine scissors to cut skull at the point of the orbital frontal suture (Red line, Fig. 1).
5. Use a scalpel to next separate the skull from spine by cutting at caudal aspect of the cerebellum.
6. Insert fine scissors into the base of the Foramen Magnum (*see* **Note 1**). Cut the skull bone at the base of the brain on the left and right side to enable the two hemispheres of the skull to flex at these points.
7. Prepare to make an incision on the dorsal surface of the skull by gathering the resected skin (**step 3**) at the base of the skull.

Firmly grasp the skin, then use fine scissors make one long cut along the sagittal suture (Green line, Fig. 1) to produce two skull halves that can be peeled away to expose brain (*see* **Note 2**).

8. Once the skull has been opened completely, place the tips of the scissors perpendicular to the long axis of the brain, and lift the separated hemispheres of the skull to enable removal of the brain below (*see* **Note 3**).
9. Rotate the head so that the dorsal surface is downward, then carefully remove brain by first dipping a spatula in PBS (so as to wet it), then "cutting" away the connecting nerves at the caudal portion of the brain, and the optic chiasm near the rostral portion, push the brain away from the skull into a petri dish or similar (sterile) nonadhesive substrate such as glass (*see* **Note 4**).

3.2 Tissue Preparation for Sectioning

1. In a 10 cm Petri dish have a small Kim wipe folded in half. Onto this Kim wipe pour 15–20 ml of the ice slurry cutting solution. Place brain onto the wet Kim wipe.
2. Place brain with the dorsal surface facing upwards and make sure olfactory bulbs are pushed together and even.
3. Use a scalpel to remove the cerebellum from the rest of the brain by making a transverse cut with a new scalpel blade immediately caudal to the cerebral hemispheres (*see* **Note 5**).
4. Place an edge of the wet Kim wipe over the brain to ensure brain does not dry out while preparing the cutting stage onto which the brain will be placed.
5. Place a drop of Super Glue GEL on the precooled (−20 °C for 10 min) vibratome cutting stage. Ensure that the Super Glue gel is spread enough to hold the brain.
6. Using forceps gently pick up the brain and blot the cut (caudal) surface of the brain on a fresh, dry Kim wipe. Too much liquid from the cutting solution will make attaching the brain to the cutting stage difficult.
7. Using forceps, gently place brain on cutting stage with cut (caudal) surface on the droplet of Super Glue Gel. The dorsal surface of the brain should be facing the front of the cutting stage.
8. Ensure brain is upright and straight (*see* **Note 6**).
9. Change gloves to minimize chance of contamination.
10. Move to vibratome.

3.3 Vibratome Sectioning of Tissue

1. Place cutting stage in vibratome that has a half double-edged razor inserted into vibratome cutting mechanism. Secure cutting stage.

2. Fill cutting stage container with ice-cold cutting slurry, use spatula to protect brain and to ensure that the brain is not dislodged during pouring (*see* **Note 7**).
3. Lower blade while moving brain backwards and forwards to enable the blade to approach the rostral tip of the olfactory bulbs.
4. Upon nicking the bulbs for the first time make note of distance on readout.
5. Move brain backward out of range of razor; lower the razor the desired depth in anticipation of performing the first cut (*see* **Note 8**).
6. Begin sectioning the brain, using slow but constant forward movement of cutting stage to ensure even cut of brain (*see* **Note 9**).
7. Continue to section the brain (discarding unwanted sections) until you reach the rostral–caudal coordinate of your region of investigation. Take careful note of the distinct landmarks for each coronal section (*see* Fig. 2).
8. Upon reaching the desired Bregma point/coronal section, use the brush to gently collect the brain section onto the spatula. Ensure that the entire brain section is flat on the spatula. Avoid folding the section.
9. Quickly transfer brain section into a small (150 μl) droplet of HEM in a large (10 cm) petri dish, which is placed on ice to keep all the sections cool. The petri dish usually has four drops of HEM spaced around the outside of the dish. A fifth droplet can be placed in the center of the petri dish. Consecutive sections are typically placed on each of the droplets in the dish in a clockwise direction to ensure sections do not become mixed up. Label the underside of the dish beforehand if you are unsure.
10. Continue cutting and collecting sections until the region of investigation has been completely obtained (*see* **Note 10**).
11. Once all samples have been collected, move petri dishes into a laminar flow hood containing your dissection microscope.

3.4 Dissection of Coronal Tissue Sections

1. Before beginning the microdissection of each coronal section, confirm that the vibratome section landmarks have been achieved (*see* **Note 11**).
2. Under a dissection microscope (×2 magnification), use fine curved forceps and scalpel (size 3 holder, size 10 blade) to dissect out the entire periventricular region (PVR) from each section. This correlates to a 3–5 cell layer-thick-region immediately adjacent to the ependymal lining of the ventricles.

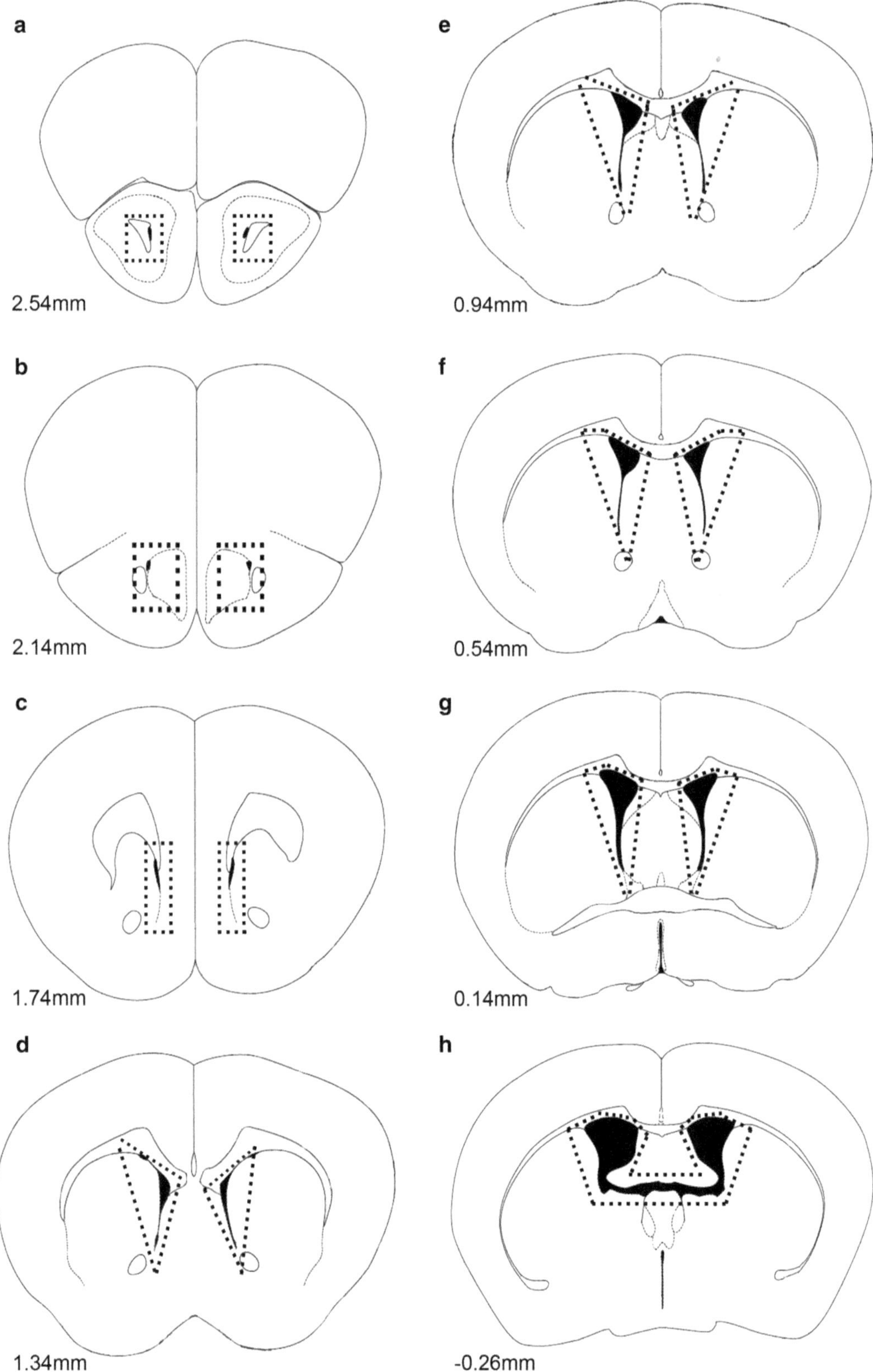

Fig. 2 Vibratome section reference. Each reference section illustrates the major landmarks at a specific

3. As the shape and location of the ventricle changes within each coronal section, the region of tissue dissected is different. The dotted lines in Fig. 2 illustrate the dissection areas for each coronal section that we commonly investigate and include as part of the LV. Minimize the amount of corpus callosum collected for each section as this can inhibit the formation of neurospheres.
4. Following dissection of the coronal section, discard unwanted parenchyma while retaining the dissected tissue within the droplet of media until all sections have been dissected. This is to avoid the tissue drying out during the course of dissection (*see* **Note 12**).
5. Once every section has been dissected use fine-curved forceps to transfer tissue of interest to a fresh petri dish. Ensure that a small amount of HEM (100 μl) is included for each section to prevent drying out. Be sure to transfer tissues in the same order as the coronal sections to ensure tissue is not confused (clockwise around the edge of the petri dish). Typically four droplets of tissue are placed into each petri dish. For ease of dicing tissue (see step below) it is recommended that the tissue be placed on the upturned lid of the petri dish.
6. Dice harvested tissue for 1 min using a fresh scalpel. Ensure all tissue on the blade is returned to petri dish by using curved forceps to clean the blade after each dissociation.
7. Wipe down forceps and scalpel blade before moving to next tissue portion. No not mix diced tissue unless you require one total number of neurospheres rather than a distribution of neurospheres throughout the LV.
8. When all tissue has been diced, place lid on petri dish and move to tissue culture hood to begin tissue culture techniques (for example, the neurosphere assay (NSA)).

3.5 Neurosphere Assay Setup

1. Use 1 ml of Trypsin–EDTA to collect each sample of diced tissue. By gently adding Trypsin to the diced tissue it should be possible to collect majority of tissue. Slowly add tissue and Trypsin mix into a 15 ml Falcon tube. Each Falcon tube should be labelled with the section number to avoid confusion. Use remaining Trypsin in pipette tip to collect any tissue that remains on the petri dish to ensure all tissue is collected.
2. Repeat collection of tissue with Trypsin–EDTA for each section.

Fig. 2 (continued) bregma point (e.g., **a** = 2.54 mm from bregma). Section **g** is used as a key reference point to determine the reliability and quality of your sections (*see* **Note 11**). *Dotted lines* demarcate the dissection areas for each coronal section

3. Place Falcon tubes containing tissue and Trypsin–EDTA into a water bath set at 37 °C for 4 min.
4. Upon removal of Falcon tubes spray with 70 % ethanol to ensure minimum risk of contamination.
5. Add 1 ml of Trypsin inhibitor to each Falcon tube and gently mix.
6. Place Falcon tubes in centrifuge for 5 min at 110 × *g*.
7. Gently remove supernatant being very careful not to disturb the pellet.
8. Resuspend pellet with 1 ml of NSA media.
9. Triturate the tissue for each section collected. This involves using a Gilson 1000 (or similar) pipette and a wetted 1,000 μl pipette tip and reducing the flow of tissue into the pipette tip by approximately 50 %. The resulting mechanical force will cause the tissue to be broken up. Trituration should continue until large chunks of tissue are no longer visible. A milky appearance of the sample is a good indicator that tissue has been adequately triturated. Care must be taken not to over triturate the tissue as this will cause the neurospheres to differentiate rather than proliferate.
10. Bring resulting tissue suspension to a final volume of 14 ml using NSA media.
11. Pass suspension through a 70 μm sieve into a fresh 15 ml Falcon tube.
12. Repeat **steps 9–11** for each section of interest that has been collected.
13. Pellet cell suspension via centrifugation for 5 min at 110 × *g*.
14. Gently remove supernatant being careful to avoid disturbing pellet.
15. Resuspend pellet using complete NSA media (containing E + F) and bring total volume for each tube to 1 ml.
16. Take 10 μl of cell suspension and mix with 10 μl Trypan blue in a microcentrifuge tube, mix and add 10 μl of this mix to a hemocytometer for a cell count.
17. Repeat **steps 15** and **16** for each section of interest collected.
18. Typically cell counts from the regions of interest (*see* Fig. 2) range from approximately 1.5 to 3.0×10^5. Each cell suspension aliquot is added to a labelled 25 cm^2 tissue flask so that total volume for each flask is 5 ml complete NSA media.
19. Flasks are then placed in a humidified incubator at 37 °C with 5 % CO_2 for 7–10 days at which point neurospheres should be ready for counting and subsequent subculturing or differentiation as desired. Neurosphere numbers are dependent upon location of the section from which the tissue was obtained.

3.6 Neural-Colony Forming Cell Assay Setup

1. As neurosphere number is dependent upon the location from which tissue was obtained we use the total neurosphere number from each Bregma point collected as the basis for plating out cells into the neural-colony forming cell assay (N-CFCA). No more than 100 colonies should be in each small (3.5 cm × 2 mm gridded petri dish). More than 100 colonies will prevent formation of large colonies that typically display neural stem cell characteristics.
2. Having obtained neurosphere numbers for the region of interest (*see* Subheading 3.4) tissue should be collected the same way as described above from Subheadings 3.1 to 3.5, **step 14**.
3. Instead of resuspending tissue in 1 ml of complete NSA medium, pellets for each region should be resuspended in 200 μl. Neurosphere number for each region will determine how many plates for the N-CFCA will be needed. For example if one region produced 500 neurospheres then 40 μl of the cell suspension should be used for each plate in the N-CFCA.

4 Notes

1. Be sure to angle the scissors downward toward the base of the skull before inserting, so as not to damage the brain above. Insert the fine scissors as deeply as possible so that 50–70 % of the skull bone will be severed.
2. Ensure upward pressure is used so the brain is not damaged.
3. To avoid damaging the brain, start at the level of the olfactory bulb, and lift the skull bone with the lower blade of the scissor before making each cut. Grasping the skin firmly will help in pulling the two hemispheres apart as you proceed.
4. Extra care needs to be taken when removing the most rostral aspect of the brain, as protective membranes surrounding the brain tend to "hold" the olfactory bulbs in place, resulting in detachment from the rest of the brain. The olfactory bulbs need to remain intact and attached for accurate vibratome sectioning.
5. Ensure cut is perpendicular both with the hemispheres and the surface of the bench (i.e., horizontally and vertically), failure to do this cut correctly will result in uneven coronal sections being obtained.
6. Tap cutting stage (with some force, not excessive) on hard surface ×3 to ensure brain is firmly affixed to cutting stage.
7. The cutting solution must be an even slush, not ice-cold liquid and not completely frozen, large pieces of icy cutting solution can interfere with the brain and the razor blade generating uneven cutting.

8. We typically section the adult brain at 400 μm, as sections ≤200 μm in thickness are difficult to obtain consistent sections, whereas ≥500 μm results in sections too thick to accurately follow and dissect the lateral ventricle due to the changing shape of this structure throughout the brain.
9. When the brain section is nearly cut all the way through, use fine bristled brush to add a slight pressure to brain, to ensure even cutting of sections.
10. It is essential to ensure that the distinct landmark(s) illustrated in Fig. 2 are used, as this will ensure that brain-to-brain variations are minimized and like regions are obtained.
11. We use Bregma 0.14 mm (Fig. 2g) as a crucial landmark to ensure that the brain has been cut evenly in a coronal orientation. By this, the anterior commissure (AC) should be clearly visible and even across both hemispheres of the tissue section. If only one hemisphere has the anterior commissure the brain has been sectioned unevenly. Brains that do not display a clear and even AC across both hemispheres in one tissue section are discarded.
12. Dissection consistency within sections and between animals is essential to ensure accurate and comparable results. The dissected regions from each hemisphere should appear the same size and shape. This will ensure consistent cell counts for each of these sections, if the sphere forming frequency/cell number is required for your experiment.

References

1. Reynolds BA, Weiss S (1992) Generation of neurons and astrocytes from isolated cells of the adult mammalian central nervous system. Science 255:1707–10
2. Gritti A et al (2002) Multipotent neural stem cells reside into the rostral extension and olfactory bulb of adult rodents. J Neurosci 22:437–45
3. Tropepe V et al (1997) Transforming growth factor-alpha null and senescent mice show decreased neural progenitor cell proliferation in the forebrain subependyma. J Neurosci 17:7850–9
4. Doetsch F et al (1999) Subventricular zone astrocytes are neural stem cells in the adult mammalian brain. Cell 97:703–16
5. Enwere E et al (2004) Aging results in reduced epidermal growth factor receptor signaling, diminished olfactory neurogenesis, and deficits in fine olfactory discrimination. J Neurosci 24:8354–65
6. Reynolds BA, Rietze RL (2005) Neural stem cells and neurospheres–re-evaluating the relationship. Nat Methods 2:333–6
7. Louis SA et al (2008) Enumeration of neural stem and progenitor cells in the neural colony forming cell assay. Stem Cells 26(4):988–96
8. Golmohammadi MG et al (2008) Comparative Analysis of the Frequency and Distribution of Stem and Progenitor Cells in the Adult Mouse Brain. Stem Cells 26(4):979–87
9. Ma Y et al (2005) A three-dimensional digital atlas database of the adult C57BL/6J mouse brain by magnetic resonance microscopy. Neuroscience 135:1203–15
10. Paxinos G, Franklin KBJ (2001) The mouse brain in stereotaxic coordinates, 2nd edn. Academic, San Diego

Chapter 17

Identifying Neural Progenitor Cells in the Adult Human Brain

Thomas I.H. Park, Henry J. Waldvogel, Johanna M. Montgomery, Edward W. Mee, Peter S. Bergin, Richard L.M. Faull, Mike Dragunow, and Maurice A. Curtis

Abstract

The discovery, in 1998, that the adult human brain contains at least two populations of progenitor cells and that progenitor cells are upregulated in response to a range of degenerative brain diseases has raised hopes for their use in replacing dying brain cells. Since these early findings the race has been on to understand the biology of progenitor cells in the human brain and they have now been isolated and studied in many major neurodegenerative diseases. Before these cells can be exploited for cell replacement purposes it is important to understand how to: (1) find them, (2) label them, (3) determine what receptors they express, (4) isolate them, and (5) examine their electrophysiological properties when differentiated. In this chapter we have described the methods we use for studying progenitor cells in the adult human brain and in particular the tissue processing, immunohistochemistry, autoradiography, progenitor cell culture, and electrophysiology on brain cells. The Neurological Foundation of New Zealand Human Brain Bank has been receiving human tissue for approximately 20 years during which time we have developed a number of unique ways to examine and isolate progenitor cells from resected surgical specimens as well as from postmortem brain tissue. There are ethical and technical considerations that are unique to working with human brain tissue and these, as well as the processing of this tissue and the culturing of it for the purpose of studying progenitor cells, are the topic of this chapter.

Key words Human brain, Neural progenitor cell, Neurosphere, Brain bank

1 Introduction

Identifying, labelling, and tracking progenitor cell birth, maturation, integration, and function in the adult human brain have been a long-held goal in neuroscience. However, ethical constraints, lack of available human tissue and scarcity of labelling techniques have all been roadblocks along the way. The first significant studies to identify progenitor cells in the brain were by two clinicians in 1944, Globus and Kuhlenbeck, who using basic histological stains on surgically resected specimens were able, by careful observation, to

Brent A. Reynolds and Loic P. Deleyrolle (eds.), *Neural Progenitor Cells: Methods and Protocols*, Methods in Molecular Biology, vol. 1059, DOI 10.1007/978-1-62703-574-3_17, © Springer Science+Business Media New York 2013

identify the stages of progenitor cell development in the human subventricular zone (SVZ). In their first description of progenitor cells, Globus and Kuhlenbeck referred to these cells as bipotential mother cells and described them as small round cells that had significant replication and differentiation abilities [1]. Between 1944 and 1998 no studies were published in which human neural progenitor cells were studied, labelled, or manipulated. In 1998, Eriksson and colleagues demonstrated the first evidence proving that the human hippocampus had the capacity for new neuron production in adulthood. Their findings were the result of obtaining the brains' of patients who had received the thymidine analogue, 5′ bromo 2′ deoxyuridine, as a part of an unrelated study on tumor spread. This single paper was the key that unlocked the door to more than a decade of intensive research on the human neural progenitor cells [2]. Today there are a myriad of published studies investigating the role and function of neural progenitors and their role in normal brain health is becoming established. The overarching hope is that human neural progenitor cells will one day be harnessed and used as a treatment for neurodegenerative diseases but even understanding their importance for the maintenance of healthy brain circuitry would be a significant step forward.

Our laboratories have an interest in the human neurodegenerative diseases and how neural precursor cells are driven to replicate and mature to compensate for the neuronal loss in conditions like Huntington's, Parkinson's, and Alzheimer's disease as well as in epilepsy and stroke. In addition we have an interest in growing and studying cultured human stem and progenitor cells for the purpose of testing novel and known compounds on human brain cells. In vitro cultures are an integral part of identifying the plasticity present in neural stem/progenitor cells and in fact the first indisputable evidence for the presence of adult mammalian neural stem cells (aNSCs) came from in vitro studies showing extensive proliferation of aNSCs from the adult rodent brain [3]. Well established in the rodents, aNSCs in culture undergo extensive self-renewal and differentiate into different cells of the neuroectodermal lineage. They can be propagated as three-dimensional sphere-like structures, known as neurospheres (NSs), and examined for their stem cell-like clonal ability through various NS assays [4, 5]. Adult human NSCs (AhNSCs) have also been successfully cultured [6–11], however, due to the difficulties in obtaining viable adult human brain tissue, in vitro culture systems for the identification of AhNSCs is yet to be fully established. Protocols and reagents are still largely based on rodent studies, and characterization studies are limited.

In this chapter we describe the currently available protocols for the isolation, maintenance, and differentiation of the AhNSCs and their characterization. Special emphasis will be placed on correctly identifying the NSC/NPCs isolated from both biopsy and autopsy brain tissues. Furthermore, electrophysiological validation of the neuronally differentiated AhNPCs will also be discussed.

The use of *postmortem* human brain tissue is subject to issues not encountered with animal studies. These include firstly, the ethical consent and cultural issues associated with obtaining tissue for research purposes, and second, the preservation of the *postmortem* human brain tissue to retain its physical and chemical integrity for rigorous scientific study (*see* **Note 1**). The human tissue that we receive is stored and managed through the Neurological Foundation of New Zealand Human Brain Bank, established by Professor RLM Faull. This tissue has been used by our own research groups and in collaborative studies with leading international research laboratories for autoradiography, in situ hybridization, electron microscopy, immunohistochemistry, progenitor cell culture, and electrophysiological validation to confirm neuronal differentiation. It is the collection, preparation, storage, and utilization of postmortem and surgical tissue specimens from the human brain for the purpose of studying neural progenitor cells that forms the basis of this chapter.

2 Materials

2.1 Brain Fixation

1. PBS-azide solution (0.1 % sodium azide in PBS).
2. 15 % Formalin (Scharlau, Spain) in 0.1 M phosphate buffer.
3. 0.1 M phosphate-buffered saline buffer (PBS pH 7.4) (KH_2PO_4, Na_2HPO_4, NaCl, KCl).
4. Sodium nitrite (BDH, UK).
5. Sucrose (BDH, UK).
6. Cryoprotectant solution: 20 % sucrose or 30 % sucrose in 0.1 M phosphate buffer with 0.1 % Na-azide (Serva, Germany). CAUTION: Na-azide is toxic.
7. Formalin pump.
8. Winged infusion needles (21 gauge –3/4 in.) attached to plastic IV hoses.

2.2 Immunohistochemistry

1. Triton X-100 (Ajax Chemicals, NZ).
2. PBS–triton buffer (0.2 % Triton in PBS).
3. DAB solution (0.05 % 3,3-diaminobenzidine tetrahydrochloride, DAB, Sigma) and 0.01 % H_2O_2 in 0.1 M phosphate buffer, pH 7.4).
4. Ni/DAB solution (1 % Ni-ammonium sulfate in DAB solution).
5. Stock 0.4 M phosphate buffer (Na_2HPO_4; NaH_2PO_4).
6. Ethanol (Andrews, local supplier).

7. Methanol (Sharlau, Spain).
8. Xylene (Andrews, local supplier).
9. DEPEX (BDH, UK).
10. Gelatine-chrome alum solution to coat glass slides (3 g gelatine and 0.3 g chromic potassium sulfate in 600 mL H_2O).
11. OCT adhesive compound (Tissue-Tek, USA).
12. Primary and secondary antibodies.

2.3 Electron Microscopy

1. Osmium tetroxide (Probing and Structure, Australia).
2. Epon (TAAB resin, UK).

2.4 Autoradiography

1. [^{3}H]-sensitive Hyperfilm (Amersham).
2. Tris–HCl buffer.
3. D19 film developer (Kodak).

2.5 Terminal Deoxynucleotidyltransferase-Mediated dATP Nick End Labeling (TUNEL)

1. 4 % paraformaldehyde in PO_4 buffer.
2. Biotin-14 dATP.
3. 1× TdT buffer.
4. 2 % bovine serum albumin.
5. 3,3-diaminobenzidine.

2.6 Equipment

1. Freezing microtome with a freezing stage (Zeiss, Germany).
2. Oxford vibratome (Oxford Instruments, UK).
3. Ultramicrotome (Leica).
4. Upright microscope (e.g., Nikon E800).
5. Laser scanning confocal microscope equipped with UV, argon, argon/krypton, and helium/neon lasers (e.g., Leica 4D or Zeiss 510 Meta).
6. Transmission electron microscope (Hitachi).
7. Digital camera (e.g., DMX1200 (Nikon) or ORCA (Hamamatsu).
8. High-power PC and imaging and quantification software (e.g., Adobe Photoshop, Adobe Systems, Inc.).

2.7 Isolation of Neural Progenitor Cells from Adult Human Brain Tissue

1. Sterile Hibernate A media (Invitrogen).
2. Sterile Hank's balanced salt solution (HBSS, Invitrogen).
3. Sterile Earle's balanced salt solution (EBSS, Invitrogen).
4. Sterile petri-dish (Nunc).
5. Sterile surgical blades (No. 23, Swann-Morton).
6. Sterile blade holder (Size 4, Swann-Morton).
7. Papain (Worthington) dissolved in EBSS to a concentration of approximately 25 units/mL. Dissolve just prior to cell isolation.

8. DNase I (Invitrogen)—Dilute in sterile cell culture grade water to a concentration of 1,000 units/mL.
9. Serological pipette (BD Falcon), 2, 5, 10, and 25 mL.
10. 70 μm cell strainer (BD Falcon).
11. Falcon tube rotator (Model no. MX001, Miltenyl Biotech).
12. Sterile Falcon tubes (15 and 50 mL; BD Falcon).
13. T75 and/or T25 tissue culture flask (Nunc).
14. 96-well cell culture plate (Nunc).
15. Adult human neural progenitor cell culture medium (Subheading 2.8.1).

2.8 Growth and Proliferation of Neural Progenitor Cells from the Adult Human Brain

1. Growth medium (Subheading 2.8.1).
2. T25 and/or T75 tissue culture flask (Nunc).
3. Accutase® (Invitrogen).
4. Sterile serological pipette (BD Falcon).
5. Sterile Falcon tubes (BD Falcon).

2.8.1 Growth Medium

Several serum-free culture mediums have been trialled. One is a commonly published medium for adult human neural progenitor cell (AhNPC) cultures and two others are commercial kits optimized for human neural stem cell research. These three medium compositions all result in successful NPC cultures; however, there are slight differences in the nature of the cellular compositions between the three mediums and this is discussed below. One should choose the medium composition for their desired purpose(s).

Note: All growth medium contained final concentration of the following, unless specified otherwise:

1. 1 % (v/v) GlutaMAX® (Invitrogen).
2. 1 % (v/v) Penicillin–streptomycin (Invitrogen).
3. 40 ng/mL Basic fibroblast growth factor (FGF-2, Peprotech).
4. 40 ng/mL Epidermal growth factor (EGF, Peprotech).
5. 2 μg/mL Heparin (Sigma).

Base medium compositions:

1. DMEM:B27 medium (DMEM:B27).
   Dulbecco's Modified Eagle Medium/Nutrient Mix F-12 (DMEM/F12; Invitrogen) supplemented with 2 % (v/v) B27 supplement (Invitrogen) (*see* Tables 1 and 2).
   (a) Most widely published for AhNPC cultures.
   (b) Supports the survival and/or proliferation of all the cell types found in our primary cultures. Primary neurons, astrocytes, microglia, as well as AhNPCs and fibroblast-like cells.

Table 1
Summary of the three growth medium compositions that can be utilized for the culture of AhNPCs

Medium	NSC/NPC selectivity	Overall cell viability	Heterogeneity	Cost
DMEM:B27	Medium	Medium	Medium	Lowest
StemPro	Highest	Lowest	Lowest	Medium
NC proliferation	Lowest	Highest	Highest	Highest

Table 2
Summary of the three differentiation medium compositions that can be utilized for the differentiation of AhNPCs

Medium	Neuronal differentiation	Overall cell viability	Heterogeneity	Cost
DMEM:1 % FBS	Yes	High	High	Lowest
DMEM:0 % FBS	Yes	Lowest	Lowest	Medium
NC proliferation	Yes (only ICC)	High	High	Highest

(c) Generates a half adherent, half suspension culture composition when plated onto non-coated tissue culture flask (T25 or T75 Nunc).

2. StemPro Medium (StemPro).
KNOCK-OUT DMEM/F12 medium supplemented with 2 % (v/v) StemPro® Neural Supplement (Invitrogen).
   (a) Developed for the growth and expansion of embryonic human neural stem cells.
   (b) Supports the maintenance and growth of all cell types, but are more selective for neurosphere forming cells compared to DMEM:B27.
3. Neurocult Proliferation medium (NC Proliferation).
NeuroCult® NS-A Basal Medium (Human, StemCell Technologies) supplemented with 10 % (v/v) of NeuroCult® NS-A Proliferation supplements (Human, StemCell Technologies).
   (a) Developed for the growth and expansion of embryonic and fetal human neural stem cells.
   (b) Supports the maintenance and growth of all primary cell types and usually results in the greatest degree of cell survival.
   (c) However, it is the least selective for NPCs, therefore results in the most heterogeneous culture out of the three types of mediums.

2.9 Differentiation of AhNPCs into Cells of the Central Nervous System

Again, there are several medium compositions that could be used to neuronally differentiate AhNPCs.

2.9.1 Differentiation Medium

All differentiation medium contained final concentration of the following, unless specified otherwise:

1. 1 % (v/v) GlutaMAX® (Invitrogen).
2. 1 % (v/v) Penicillin–streptomycin (Invitrogen).
3. 40 ng/mL Nerve growth factor (NGF, Peprotech).
4. 40 ng/mL Neurotrophic factor-3 (NT-3, Peprotech).
5. 40 ng/mL Brain derived neurotrophic factor (BDNF, Peprotech).

Base medium compositions:

1. DMEM:F12 + 1 % FBS medium (DMEM:1 % FBS).
   Dulbecco's Modified Eagle Medium/Nutrient Mix F-12 (DMEM/F12; Invitrogen) supplemented with 1 % (v/v) fetal bovine serum (FBS, Invitrogen).
   (a) Widely used to differentiate NSCs/NPCs into neurons and astrocytes.
   (b) Results in physiologically active neurons and astrocytes.
   (c) However, also supports the growth of fibroblast-like cells, and in long-term cultures, the fibroblast-like cells will dominate the culture.
2. DMEM:F12 + No serum medium (DMEM:0 % FBS).
   Dulbecco's Modified Eagle Medium/Nutrient Mix F-12 (DMEM/F12; Invitrogen) supplemented with 2 % (v/v) B27 (Invitrogen).
   (a) A no serum version of the medium composition above, which also results in electrically active neurons and astrocytes.
   (b) Limits the growth and proliferation of the fibroblast-like cells.
   (c) However, there is a slight compromise in cell viability, when compared to the DMEM:1 % medium.
3. NeuroCult Differentiation medium (NC Differentiation).
   NeuroCult® NS-A Basal Medium (Human, StemCell Technologies) supplemented with 10 % (v/v) of NeuroCult® NS-A Differentiation supplements (Human, StemCell Technologies).
   (a) Commercially available neuronal differentiation medium designed to differentiate embryonic and fetal NSCs/NPCs.
   (b) Results in antigenitically identifiable neurons and astrocytes (electrophysiological confirmation pending).
   (c) Similar culture patterns as those seen with DMEM:1 % FBS medium, as it seems to also support the fibroblast-like cell population.

2.10 Electrophysiological Analysis of the Differentiated AhNPCs

1. External (artificial cerebral spinal fluid; ACSF) solution (Table 3).
2. Internal patch pipette solution (Table 4).
3. 12 mm, #1 thickness, round coverslip (Menzel-Glaser).
4. MilliQ water.
5. Upright microscope with a recording chamber (Olympus BX51W1).
6. Carbogen (BOC–Local supplier).
7. Multiclamp 700B amplifier (Axon instruments).
8. pClamp 8 software (Axon instruments).
9. Electronic pipette manipulator (Sutter Instruments).
10. Digitizer (Sutter Instruments).
11. Fire polished 1.15 mm inner diameter borosilicate glass (King Precision Glass, Inc.)
12. Pipette puller (Sutter Instruments).

Table 3
Chemical composition of artificial cerebral spinal fluid (external solution)

Chemicals	Final concentration (mM)
NaCl	150
KCl	4
$CaCl_2$	2
$MgCl_2$	2
HEPES	10
Glucose (added prior to experiment)	10

Table 4
Chemical compositions of patch pipette internal solution

Chemicals	Final concentration (mM)
K^+ gluconate	120
HEPES	40
$MgCl_2$	5
NaATP	2
NaGTP	0.3

3 Methods

3.1 Immunohistochemical Methodologies

3.1.1 Processing of Postmortem Human Brain Tissue

Brain tissue is obtained for the brain bank in three ways:

1. Diseased brains from cases that died from a neurodegenerative disease in which we have a research interest are obtained through a donor program where fully informed consent has been granted by the patient and/or the family in advance of the person's death.
2. Normal brains, that is, brains from cases that have died with no history of any neurological or psychiatric disorder are obtained through a tissue donor coordinator at the hospital mortuary where full consent is obtained from the family.
3. Intraoperative specimens are obtained from patients undergoing surgical resections, usually of the temporal lobe and hippocampus as a treatment for epilepsy.

In our laboratory we use two methods of preparing the brains. We either (1) perfuse the entire brain with formalin and then block it or (2) separate the hemispheres and perfuse one hemisphere and block it, and the other hemisphere we freeze and block as fresh unfixed tissue blocks. This procedure of cutting the blocks offers several advantages over other methods:

1. It is designed so that specific regions located on particular gyri of the cortex, and other regions such as the basal ganglia, hippocampus, and brainstem are cut into easily identifiable blocks that can be sectioned or homogenized.
2. Because of this, it allows stereological counting methods to be used on specific brain regions. The following steps describe the fixation and dissection of *postmortem* human brain in detail, followed by the immunohistochemical protocols for use of the tissue.

Processing of Fresh (Unfixed) Postmortem Human Brain Tissue

For brains where one hemisphere is not perfused, and kept fresh, it is blocked following exactly the same procedures as for the fixed hemisphere (see below **step 2** brain dissection and Fig. 1).

1. The blocks are placed into metal dishes and frozen with powdered dry ice, double wrapped in aluminum foil, code labelled, and stored at −80 °C.
2. These fresh tissue blocks are used for biochemical/gene/protein analysis or frozen sections are cut on a cryostat and used for receptor autoradiography labelling (*see* Fig. 2).
3. In addition, tissue used for cell culture can also be isolated from this block (*see* Subheading 3.2).

The quality of the tissue is essential for this protocol. The *postmortem* delay and quick processing of the brain tissue are

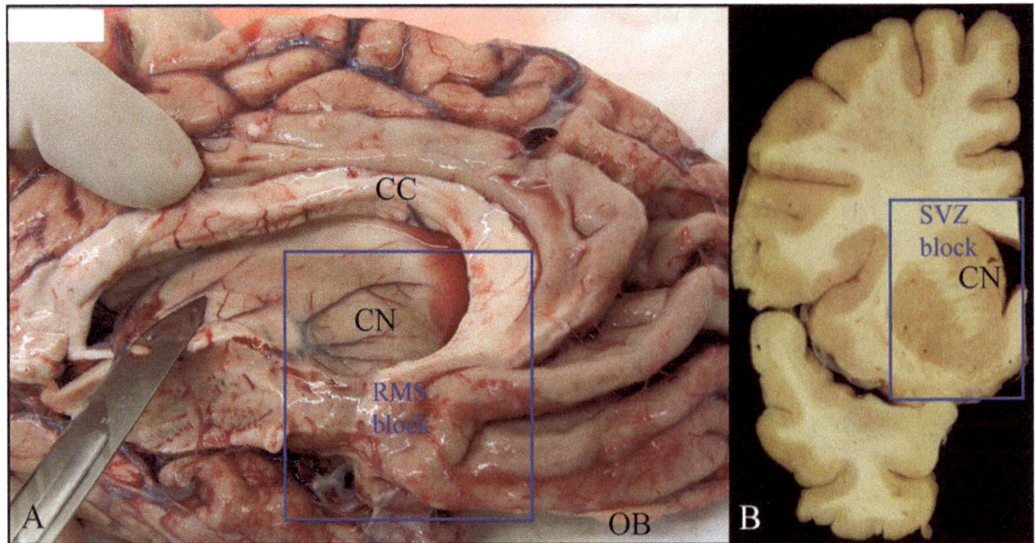

Fig. 1 (**a**) Photograph of the medial aspect of the human forebrain. The *blue box* reveals the block required to further cut and stain the rostral migratory stream (RMS). *CN* caudate nucleus, *OB* olfactory bulb, *CC* corpus callosum. (**b**) Is a photograph of the human forebrain cut coronally. The *blue box* reveals the block that is further sectioned and stained to visualize the subventricular zone (SVZ)

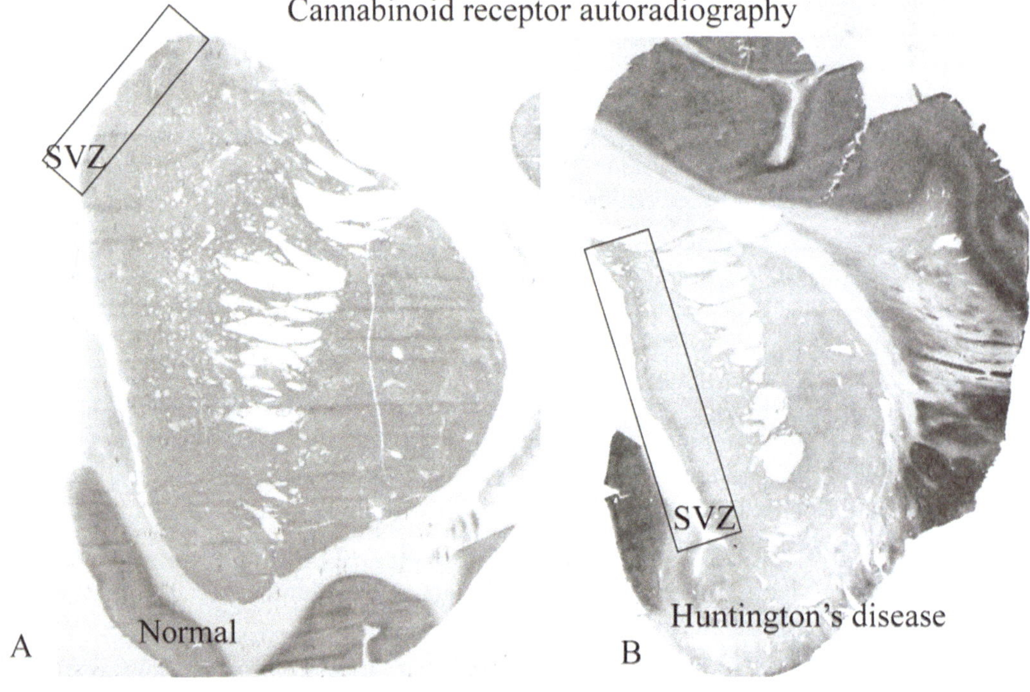

Fig. 2 (**a**) Demonstrates a photograph of the autoradiographic film produced by radioligand labelling the cannabinoid receptor Cb1 with the [3H] CP55,940 ligand. The figure shows a normal caudate nucleus in which there is homogenous binding of the radioligand across the caudate nucleus. (**b**) Reveals the same ligand but this time it is an advanced grade of Huntington's disease that is demonstrated. The results reveal that the SVZ has extensive radioligand binding despite the reduction in Cb1 binding in the adjacent caudate nucleus. These results demonstrate the suitability of autoradiography for studying receptors and ligand binding in stem cell niches in the human brain

crucial determining factors. For optimal immunohistochemical studies on fixed human tissue, only human brain tissue with a *postmortem* interval less than 24 h after death is accepted for the brain bank.

Processing of Postmortem Human Brain Tissue for Fixation in Formalin

1. On receipt, the brain is weighed and placed on a mesh metal drainage tray. If the whole brain is to be perfusion-fixed, then identify the vertebral and carotid arteries. (If one hemisphere is to be frozen and one hemisphere is to be perfusion-fixed then separate the hemispheres.)
2. Remove the meninges so that only the midbrain holds fore- and hindbrain together.
3. Make a left and right incision above the substantia nigra (SN) at the junction of the midbrain and forebrain.
4. The incision should be at 45° and should be as deep as possible. The brain stem and cerebellum can now be removed and blocked in the following way:
   (a) The cerebellum (Cb) and the brainstem are separated from the hemispheres by cutting through the three cerebral peduncles and divided into blocks of cerebellum, the midbrain, pons, and the medulla approximately 0.5–1 cm thick coronal slices.
   (b) These blocks can be immediately put into fixative or frozen as fresh tissue, as required.
5. The two hemispheres are now carefully separated by cutting through the corpus callosum (CC).
6. Make sure that the vertebral and carotid blood vessels on the side to be perfused including the circle of Willis are kept intact.
7. Fix the whole brain or single hemispheres by perfusion through the basilar and internal carotid arteries (whole brain) and through the carotid, vertebral and anterior cerebral arteries (for the half brain).
8. Insert winged infusion needles (21 gauge), attached to plastic IV hoses, into the respective arteries and tie securely. The hoses are attached to a perfusion pump through which the perfusion solutions can be pumped throughout the brain at a constant rate (approximately 2 L per hour).
9. Stop any leaks by tying the leaking vessels off.
10. The brain is first flushed with a solution of PBS (pH 7.4) with 1 % sodium nitrite to wash out the blood (approximately 15 min or until the brain is cleared of blood) and to cause mild vasodilation and thus a more global fixation, followed by 2 L of fixative over about 30–45 min.

11. The fixative consists of 15 % formalin (diluted from a stock solution of 37 % formaldehyde: 37 % formaldehyde is taken as 100 % stock solution) in 0.1 M phosphate buffer, pH 7.4.
12. After perfusion, the brain is post-fixed for 6–12 h in the same fixative before being dissected into blocks.

Dissection of the Blocks

1. Each hemisphere is laid on its medial side to expose the lateral side of the brain.
2. The various cortical regions are blocked according to their gyri (*see* protocols ref. 12) and code labeled.
3. Blocks from the same cortical region are identified by grooving (no cut, one cut, two cuts, and three cuts) into the white matter.
4. The samples are stored in labeled biohazard bags.
5. Cut the remaining brain in the coronal plane in the rostral to caudal direction, starting at the rostral pole of the caudate-putamen to the most caudate pole of the thalamus.
6. Trim the blocks to contain the various levels of the basal ganglia and thalamus in the following sequence:
   (a) The caudate nucleus and putamen (CN).
   (b) The ventral striatum (VS).
   (c) The globus pallidus (GP), globus pallidus, and anterior thalamus (GP-Thal).
   (d) Thalamus (THAL).
   (e) Dissect the amygdala (AMY) and the hippocampus (HP) from these blocks and divide it rostro-caudally into equal 1 cm blocks and label with the above abbreviations.

This is the standard method we use for brain storage and this is the most versatile way to ensure that coronal sections can be cut from blocks containing the progenitor niches, the rostral migratory stream (RMS), SVZ, hippocampus, and amygdala (*see* Fig. 1a, b). However if the brain is to be used specifically for studies of the RMS then an alternative method of dissecting is required for the block containing the CN and GP.

1. Make an incision running rostral to the CN from half way along the olfactory tract up to the corpus callosum.
2. Make another incision at the level of the anterior commisure and a superior incision between the first two incisions.
3. This block can then be removed and after further fixation and sucrose treatment can be cut sagitally to study the RMS (*see* Fig. 1a).

3.1.2 Immuno-histochemical Processing

1. Place the dissected blocks into fresh fixative for 24–48 h.
2. Transfer these blocks first to 20 % sucrose in 0.1 M phosphate buffer with 0.1 % Na-azide for 2–3 days.
3. Thereafter, change the solution over to 30 % sucrose in 0.1 M phosphate buffer with 0.1 % Na-azide and incubate for a further 2–3 days for cryoprotection.
4. After complete infiltration of the 30 % sucrose, snap-freeze tissue blocks on powdered dry ice, double wrap the blocks in aluminum foil. The blocks are coded and numbered and stored in a −80 °C freezer.

Pretreatment of the Sections

1. Select adjacent series of sections and process free-floating in tissue culture wells.
2. Wash the sections in PBS and 0.2 % Triton-X (PBS–triton) −1 h to overnight.
3. Optional: To reduce nonspecific background staining and increase labelling for certain antibodies (see technical notes) by "unmasking" of antigenic epitopes, pretreatment of the fixed tissue before processing for immunohistochemistry is carried out using the following procedure [13–17]:
   (a) The sections for antigen retrieval are transferred to six-well tissue culture plates and incubated overnight in 0.1 M sodium citrate buffer, pH 4.5 (*see* **Note 2**).
   (b) They are then transferred to 10 mL of fresh sodium citrate buffer solution per tissue culture well.
   (c) The plates are then placed in the center of a 650 W microwave oven and the sections are microwaved for 30 s and allow to cool to room temperature (note that each of the 6 wells must have 10 mL of solution in them, even if the well has no tissue in it, so a consistent boiling point is obtained across experiments).
4. Wash the sections in PBS–triton (3 × 15 min).
5. Incubate the sections for 20 min in 50 % methanol with 1 % H_2O_2 to block endogenous peroxidases in the tissue.
6. Wash the sections in PBS–triton (3 × 15 min).
7. Add the primary antibodies according to the research of interest (Table 5).

The localization of the primary antibodies can be viewed with various detection systems.

Immunoperoxidase Labeling

1. Incubate the sections in primary antibodies in immunobuffer at their appropriate dilution for 2–3 days on a shaker at 4 °C.
2. Wash the sections in PBS–triton (3 × 15 min) to wash off the primary antibodies.

Table 5
Table displaying antibodies commonly used in progenitor cell research in human brain and the cell or cellular component detected by the antibody

Antibody name	Species	Supplier	Dilution	Cell or cell component detected
β-cell lymphoma (BCL2)	Mouse	SantaCruz	1:300	Antiapoptotic factor in neurons
βIII tubulin	Mouse	Chemicon	1:500	Cytoskeletal marker of immature neurons
Bromodeoxyuridine	Mouse	Hara Sera Lab	1:500	Thymidine analog detection
Calbindin	Mouse	P. Emson	1:5,000	Neuronal calcium binding protein
Doublecortin (Dcx)	Goat/rabbit	SantaCruz	1:500	Neurofilament protein in young neurons
Ferritin	Rabbit	Sigma	1:1,000	Microglial marker
Gamma 2 (γ2)	G.Pig	H.Mohler	1:2,000	Subunit of $GABA_A$ receptor
Glial fibrillary acidic protein	Mouse	Dako	1:3,000	Fibrillary astrocyte marker
Meis-2	Mouse	Gifted	1:1,000	Transcription factor in developing medium spiny neuron
Mini chromosome maintenance protein (MCM2)	Mouse	SantaCruz	1:500	Cell cycle protein
Musashi 1 (Msi1)	Rabbit	Millipore	1:2,000	Neural RNA binding protein
Nestin	Rabbit	R.McKay	1:1,000	Intermediate filament protein found in progenitor cells
Polysialylated-neural cell adhesion molecule (PSA-NCAM)	Mouse	Chemicon	1:1,000	Migratory neuroblasts
Proliferating cell nuclear antigen (PCNA)	Mouse/rabbit	SantaCruz	1:500 1:1,000	Cell cycle protein
Vimentin	Mouse	Chemicon	1:500	Intermediate filament protein

3. Incubate the sections overnight in species-specific biotinylated secondary antibodies (various companies, we use goat-anti-mouse at 1:1,000; Sigma; Chemicon, USA) diluted in immunobuffer.
4. Next day, wash the sections in PBS–triton (3 × 15 min) to wash off the secondary antibodies.

5. Incubate the sections for 4 h at room temperature in ExtrAvidin™, (diluted 1:1,000, Sigma). Alternatively you can use an avidin–biotin–peroxidase complex kit.
6. Wash the sections in PBS–triton (3 × 15 min).
7. React the sections in DAB solution for 15–30 min to produce a brown reaction product. A nickel-intensified procedure (using Ni/DAB instead of DAB solution) can also be used to produce a blue-black reaction product [18].

Immunofluorescent Double- or Triple-Labelling

To detect multiple antigens in a single tissue section primary antibodies raised in different species (e.g., mouse, rabbit, guinea pig, or sheep) are used. The secondary antibodies (anti-mouse, anti-rabbit, anti-guinea pig, anti-sheep) are coupled to different fluorochromes (e.g., Alexa Fluor 488, Cy3, Cy5) and are best to be raised in the same species (e.g., donkey, goat). A large number of double- and triple-labelling combinations using a cocktail of monoclonal and polyclonal primary antibodies are possible followed by secondary antibodies linked to different fluorochromes.

1. Incubate the sections in primary antibodies diluted in staining buffer for 2–3 days on a shaker at 4 °C.
2. Wash the sections in PBS–triton (3 × 15 min) to wash off the primary antibodies.
3. Incubate the sections in species-specific fluorescent secondary antibodies directly linked to fluorophores diluted in immunobuffer (e.g., goat anti-rabbit AlexaFluor488 and goat anti-mouse AlexaFluor594, Molecular Probes) overnight on a shaker at 4 °C.
4. Fluorophores of various wavelengths are now available for various combinations of primary antibodies.

3.1.3 Electron Microscopy

Fix the blocks from regions of the brain that are to be investigated at the ultrastructural level and embed in resin as follows.

1. Dissect very small blocks of tissue (2 mm^3 blocks) directly from the fresh brain and immerse them in the desired fixative, either 2.5 % glutaraldehyde in 0.1 M phosphate buffer for traditional tissue ultrastructure or 4 % paraformaldehyde and 0.5 % glutaraldehyde in 0.1 M phosphate buffer for pre-embedding immunoelectron microscopy.
2. Alternatively, sections cut from perfusion fixed brains on a vibratome in 0.1 M phosphate buffer at a thickness of 100 μm and post-fixed in either of the above fixatives as required.
3. Process these sections for conventional electron microscopy or immunoelectron microscopy (*see* **Note 3**).

3.1.4 Autoradiographic Procedures

In our laboratory, unfixed frozen tissue blocks are used to localize the ligand binding sites of neurotransmitter receptors such as benzodiazepines, opiates, adenosine, cannabinoids, and others. The slide mounted sections are processed for the autoradiographic localization of receptor binding ligands in which the autoradiographically labelled receptors appear as black dots on a clear background [19–25].

1. Blocks are sectioned at 16 μm on a cryostat and the sections thaw-mounted onto gelatine/chrome alum-coated slides.
2. Incubate the cryostat sections in 50 mM Tris–HCl (pH 7.4) containing, for example, 1 nM [3H] flunitrazepam (84 Ci/mmol, Amersham).
3. Incubate the sections with a receptor binding molecule.
   (a) Example: GABAA receptor agonist with a high affinity for Type I and Type II GABAA/benzodiazepine receptors or [3H]Ro 15-1788 (82.8 Ci/mmol, New England Nuclear), a GABAA receptor antagonist with a high affinity for central Type I and Type II receptors and a. To demonstrate the binding to Type I receptors adjacent sections were incubated with [3H]flunitrazepam or [3H]Ro 15-1788 in the presence of CL 218-872 (a ligand with high affinity for Type I and a low affinity for Type II GABAA/benzodiazepine receptors). For SVZ labelling we have used cannabinoid receptor Cb1 with the [3H] CP55,940 ligand.
4. The sections are washed (2 × 1 min in Tris–HCl buffer, with a final dip in ice-cold distilled water) and dried under a stream of cold air.
5. All procedures are carried out at 4 °C.
6. Non-specific binding is determined by the incubation of slides in the presence of 1 μM clonazepam.
7. Once dry, the slides are brought to room temperature, taped into X-ray cassettes and apposed with [^{3}H]-sensitive Hyperfilm (Amersham) and exposed in the dark at 4 °C for 6–12 weeks.
8. The films are developed in D19, washed, fixed, and dried. They are printed using standard photographic procedures to yield autoradiograms in which the autoradiographically labelled receptors appear as white dots on a black background.

3.1.5 TUNEL Staining

Terminal deoxynucleotidyltransferase-mediated dATP nick end labeling (TUNEL) is an important technique for understanding progenitor cell biology. It is one way to determine whether cell death is a major outcome for a given population of progenitor cells in the brain. The principle on which it is based is that during apoptotic events the DNA within a cell is fragmented leaving an abundance of DNA ends to label that in a healthy cell would not be exposed to label. TUNEL labelling has the disadvantage of detecting the process of fragmentation for only a very short period during the process of apoptosis and thus the number of TUNEL positive cells visible at any given time is small compared with the large number that may indeed fragment their DNA. We use it in most cases where we are trying to determine the fate of the progenitor in the cell pool where it is not clear whether the progenitors are maturing or dying.

1. 16 μm cryostat cut fresh sections are mounted onto chrome alum-dipped slides, thawed slightly prior to fixation with 4 % paraformaldehyde in PO_4 buffer, placed in 1 % H_2O_2 in methanol.
2. Sections are preincubated in 1× TdT buffer before being incubated in a medium comprising TdT, 5× TdT buffer, biotin-14 dATP, and dH_2O (37 °C, 1 h).
3. Sections are rinsed in 2× standard saline citrate, dH_2O, and PBS prior to blocking in 2 % bovine serum albumin (10 min, room temp.), washed and incubated in extravidin peroxidase (1:100 in PBS, 1 h).
4. Latent staining are achieved using 3,3-diaminobenzidine and 0.01 % H_2O_2.
5. Sections are dehydrated through graded alcohols, cleared in xylene and coverslipped with hystomount.

3.2 In Vitro Culture and Identification of Adult Human Neural Progenitor Cells

3.2.1 Isolation and Enrichment of Neural Progenitor Cells from Adult Human Brain Tissue

In this section we describe the isolation and some optional enrichment procedures that can be implemented when culturing AhNPCs.

Adult human primary cell studies will result in variability and two cases will seldom be identical. Other than those arising from individual variation, the following can influence the outcome of the cultures.

1. Source of the brain tissue.
   (a) Biopsy brain tissues usually have a higher success rate compared to those from autopsy, possibly due to significantly shorter delays experienced by the tissue prior to being isolated in the laboratory.
2. Post-operation or postmortem delay.
   (a) Shorter the better. However, neural stem/progenitor cells are more resistant to metabolic stress compared to post-mitotic neurons and astrocytes, and this allows them to be reliably isolated from tissues that have 4 h post-operation or postmortem delay or less.

3. Regions of culture.
   (a) Although some NPCs may exist in the outer cortical areas, the most reliable source of NPCs is the SVZ of the lateral ventricle (this encompasses the entire length of the lateral ventricle, including the inferior horn).
   (b) Hippocampal tissue that includes the dentate gyrus can also act as a source, albeit at lower efficiencies compared to the SVZ.

Wherever possible, the removal and transport of brain tissue should be conducted aseptically. If this is not possible (i.e., many postmortem specimens are obtained from mortuaries that do not use aseptic techniques), one can incubate the tissue in nutrient medium containing antibiotic/antimycotic agents for several hours at 4 °C before proceeding to the isolation step.

Dissection of Tissue from Neurogenic Regions of the Human Brain

Biopsy tissue (*Temporal Lobectomies*). The majority of biopsy brain tissues obtained for the culture of neural progenitor cells arise from temporal lobectomy surgeries for intractable epilepsy. In most cases, the specimen will include the ventricular surface of the inferior horn of the lateral ventricle that is situated between the hippocampus and the medial surface of the temporal lobe (Fig. 3). If you cannot locate this region in your sample, only proceed with the hippocampal region containing the ventricular zone and the dentate gyrus.

1. Prepare the tissue dissociation medium by first dissolving one vial of papain in 5 mL of either Hibernate A or EBSS medium.
2. To make up 10 mL of dissociation medium add 1.25 mL of the above papain and 100 μL of DNase1. Later, 10 mL of dissociation medium will be added per gram of tissue.

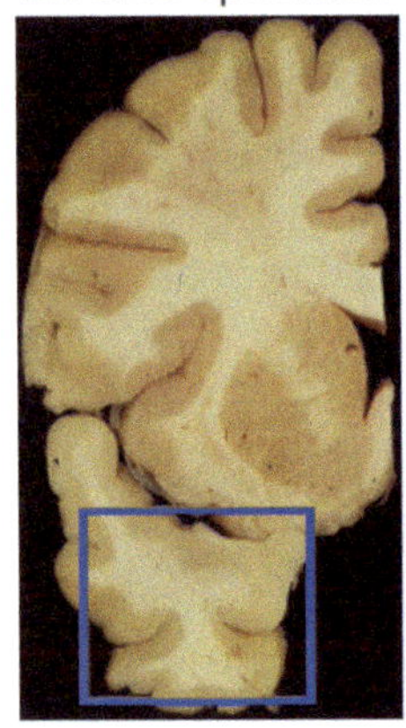

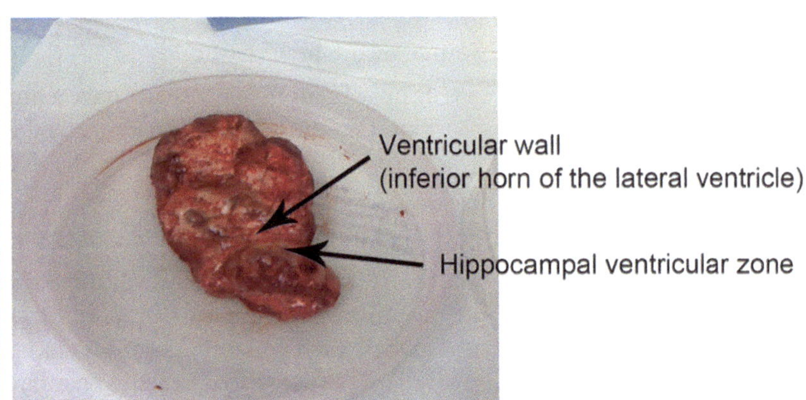

Fig. 3 Images illustrating the representative brain region obtained from majority of our biopsy tissue arising from temporal lobectomy. *Right images* shows are of an *en bloc* specimen of one such case. *Arrows* indicate the neurogenic regions of interest that are dissected and cultured from these biopsy tissue specimens

3. Warm the culture medium and the tissue dissociation medium to 37 °C in a clean water bath containing antimicrobial agents.
4. Collect tissue in a sterile container containing precooled (4 °C) hibernate A media and bring back to the lab.
5. Place the tissue onto a 100-mm sterile petri-dish in a tissue culture grade laminar flow hood (whole block if it was an *en bloc* surgery or the hippocampal region if it was a separate excision). Pour enough cold hibernate A medium to just cover the surface of the Petri-dish.
6. As shown in Fig. 3, using a sterile scalpel blade, dissect the ventricular region surrounding the hippocampus and, if visible, the ventricular zone along the medial surface of the temporal lobe. You can also pool the hippocampal DG region to increase NPC yield if discriminating between SVZ and DG is not of importance.
7. Place the tissue regions of interest into a pre-weighed falcon tube containing Hibernate A and re-weigh the sample to calculate the weight of the tissue.

Postmortem tissue. Although NPC cultures from postmortem tissues generally have a lower chance of success, there are definite advantages of utilizing this tissue source. One, all neurogenic regions of the brain are available for culture: two, more tissue can be obtained for culture: and three, NPCs can be isolated from brains with various neurological and non-neurological conditions.

1. Dissections of several well-known regions of adult neurogenesis are illustrated in Fig. 4.
2. Anterior SVZ region of the lateral ventricle can be dissected from both the coronal and sagittal sections of the brain.
   (a) In coronal sections, once the ventricular region encompassing the caudate nucleus is visible, take a 1 cm thick coronal section and dissect 0.5 cm of the ventricular lining.
   (b) In the sagittal plane, hemidissect the entire brain and open the brain up to view the medial surface. Using the caudate nucleus, locate the ventricular region by identifying the indented medial portion of the brain that is shinny with a slightly dark tint (caused by the caudate sitting behind it). Using the thalamostriate vein as the caudal guideline, cut a rectangular block from this region to the rostral portion of the lateral ventricle.
3. The inferior horn of the lateral ventricle and the hippocampus can be isolated using an *en bloc* style surgical incision approach of the middle to caudal temporal lobe areas (Fig. 4). Using a scalpel, make a deep incision into the superior temporal lobe until the ventricular region is visible (evident by the rich

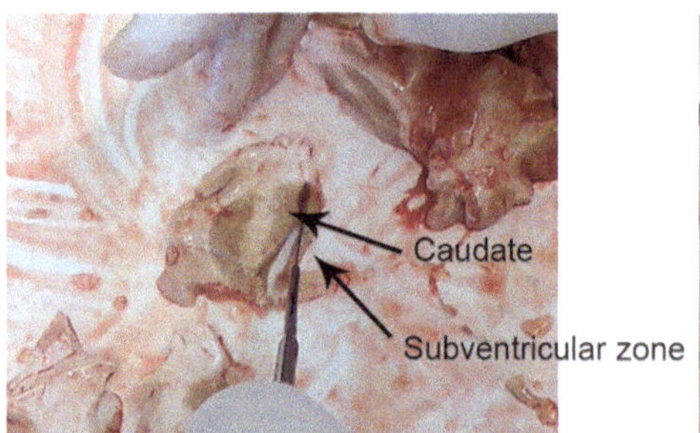

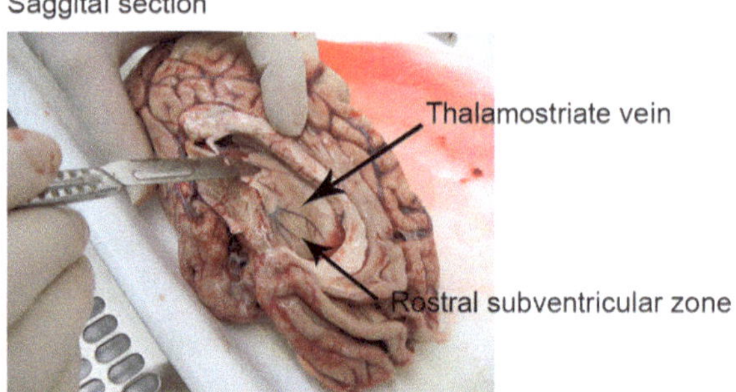

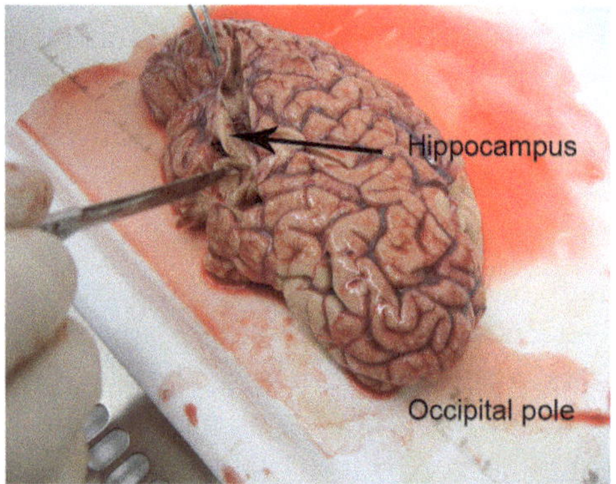

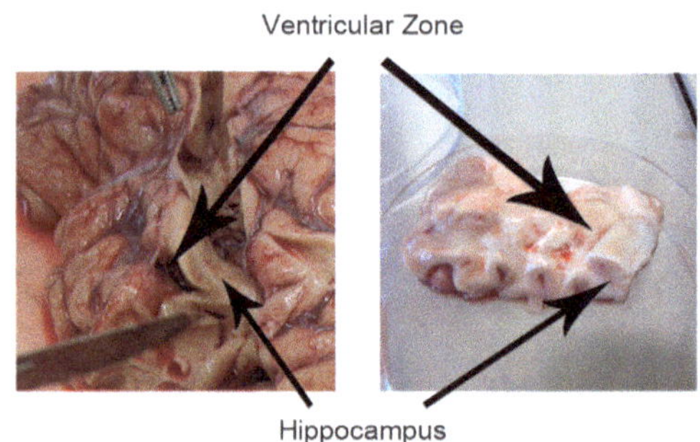

Fig. 4 Images illustrating the regions dissected for in vitro culture of AhNPCs from the postmortem adult human brain. The *top row* shows the dissection of the rostral region of the SVZ adjacent to the caudate nucleus and the lateral ventricle. This region is a major neurogenic area in the adult human brain and gives rise to the rostral migratory stream. *Top left* represents a coronal dissection and the *top right* represents a dissection through the medial surface of the lateral ventricle. *Bottom row* of images shows the dissection of the hippocampus along with the SVZ lining the inferior horn of the lateral ventricle. An *en bloc* type of incision is made through the temporal lobe and the block is excised out. The medial surface of this block will contain both the hippocampus and the ventricular zones

vascular structure of the choroid plexus). From the boundaries of that cut, make two vertical incisions down the temporal lobe to completely remove the temporal block. Turn the block to its medial surface and remove the remaining choroid plexus to locate the ventricular surface. The hippocampus can be identified by the bulge in the medial surface of the temporal block, as well as the characteristic dentate gyrus, visible when viewed from the coronal plain. Dissect 0.5 cm deep sections along the ventricle, which will include the hippocampal ventricular lining. Cut deeper into the hippocampus to include the dentate gyrus.

Dissociation and Initial Culture of AhNPCs

1. Using two scalpel blades, dice the tissue into <2 mm^2 sections to maximize surface area for efficient enzymatic digestion. Do not take longer than 5 min on this step.
2. Using a 5 mL serological pipette, transfer the diced tissue into a 50 mL falcon tube with 20 mL of Hibernate A medium and let the tissue sections settle.
3. Carefully remove the floating debris by aspirating most of the media and adding designated amount (10 mL per 1 g of tissue) of pre-warmed dissociation media.
4. Transfer the 50 mL falcon tube containing tissue/enzymatic mix onto a rotor and allow for gentle rotation (12 rpm) for 10 min at 37 °C.
5. After the first 10 min, undergo gentle trituration using a 10 mL serological pipette to further break up the tissue and continue with the enzymatic digestion at 37 °C on a rotor for a further 10 min.
6. Slow the digestion process by adding an equal volume of DMEM:F12 B27 media (without growth factors).
7. Triturate sample until no tissue clumps are visible (typically 10 triturations using a 10 mL serological pipette).
8. Transfer the above dissociation mixture into a new 50 mL falcon tube through a 70 μm cell strainer to filter out undigested tissue and debris.
9. Centrifuge the filtrate at 170 × *g* for 7 min.
10. Remove all the supernatant and resuspend the cells in AhNPC maintenance of choice.
11. Plate 1 g of tissue per T25 tissue culture flask in 6 mL of medium.
12. Leave for 12–24 h in a tissue culture incubator (37 °C, 5 % CO_2).

Optional: Erythrocyte lysis step to remove red blood cell contamination. This significantly aids counting of the cells and may help the future viability of the NPCs.

1. After **step 9** above, resuspend the cells in 5 mL of the buffer composition stated below.
   (a) Erythrocyte lysis buffer = 155 mM NH_4Cl, 1 mM $KHCO_3$, 0.1 mM EDTA and pH 8.0.
2. Incubate on ice for 15 min.
3. Add 20 mL of DMEM:F12 B27 media (without growth factors) and centrifuge at 170 × *g* for 5 min.
4. Continue with **steps 10** above.

Enrichment of AhNPC Cultures

The initial cultures (especially from biopsy cases) generally contain a heterogeneous cell population comprised of AhNPCs, microglia, neurons, astrocytes and fibroblast-like cells. If this is not desirable or if you would like to separate certain cell populations, there are couple of methods that utilize the fact that certain cell types adhere faster and better to the culture flask substrate than the NPCs, and that some do not proliferate in culture. The widely used FACS and magnetic sorting is yet optimised for AhNPC cultures, due to the fact that our yields are low and there are no reliable cell surface antigens to accurately select neurons and astrocytes. Reporter genes have been used, however, these require transfections or transductions, both of which also decrease cell viability.

Microglial isolation. Microglial cells strongly adhere to the culture surface within a couple of hours of plating. Therefore, replating the non/less adherent cells after 6–12 h post-isolation is a convenient way of removing the majority of microglia.

1. After 6–12 h post-isolation, gently tap the tissue culture flask to suspend all non-adherent or less adherent cells into the culture medium (Microglial cells should remain adherent to the culture flask).
2. Remove the entire culture medium and centrifuge at $170 \times g$ for 5 min.
3. Meanwhile, add fresh AhNPC proliferation medium to the flask containing the adherent cells. These cells can be further cultured and used for microglial experimentation.
4. After centrifugation, dispose all the supernatant and resuspend in equal volume of fresh AhNPC proliferation medium and transfer to a new tissue culture flask.

Post-mitotic cell population isolation (only biopsy cultures). From biopsy specimens, we generally obtain a large number of astrocytes (and to a lesser extent, neurons). These are difficult to select based on adherent characteristics, therefore we rely on their post-mitotic nature to dilute them out from culture over time.

1. The initial passages (P1-2) will contain a large number of astrocytes and some neurons. Experiments designed to utilize these primary cells should be conducted during this time frame.
2. The post-mitotic cells dramatically decrease in number and by passage 3.

Mitotic cell population. As the post-mitotic cells get diluted out, only the mitotic cell populations, AhNPCs and the fibroblast-like cells [26], should remain in culture The fibroblast-like cells have high proliferative capacity and we advise readers to conduct AhNPC-based experiments between passages 2–5.

3.3 Maintenance of AhNPC Cultures

After the initial replate, half change the proliferation media every second or third day.

1. Using a serological pipette, carefully remove half the media from each culture flask into a separate falcon tube.
2. Centrifuge the falcon tube at 50 × *g* for 3 min to collect the non-adherent cells/spheres.
3. Carefully aspirate the supernatant and resuspend in equal volume of fresh proliferation medium. Carefully resuspend the cells using a serological pipette.

   Note: Have the tip of the pipette just above the bottom of the tube (not pressed against it), to ensure good resuspension but minimal sphere breakup.
4. Place the fresh media, plus cells, back into the original culture flask to complete the half media change.

3.4 Passaging of AhNPCs

Once the majority of the spheres reach approximately 300 μm in diameter (or their centers start to appear dark/pigmented), and/or the monolayer of cells reach 90 % confluence, proceed with passaging. We have found single cell suspension generations to be detrimental to the neurospheres/NPC cultures; therefore, we employ a less vigorous method that only breaks up larger spheres into smaller ones to avoid single cell suspensions.

As most cultures will contain a mixture of adherent and non-adherent cells/spheres, a two-step process must be employed.

1. First, gently tap the culture flask to loosen the non/semi-adherent cells/spheres and fully remove the media and place it in a falcon tube.
2. Wash the flask again with half the volume of fresh proliferation medium and add this to the same falcon tube.
3. Centrifuge the falcon tube at 170 × *g* for 3 min.
4. Meanwhile, add pre-warmed Accutase® to the culture flask to detach the adherent cells and incubate in a 37 °C incubator for approximately 5 min.
5. Centrifugation would have completed by this time. Remove the supernatant and keep it as conditioned media in a separate falcon tube.
6. Once the adherent cells have been detached, remove all the Accutase® plus cells and add them directly into the falcon tube with the pelleted non-adherent cells. Resuspend using a serological pipette and triturate the mixture 10 times before placing it in a 37 °C water bath for 2 min.
7. This procedure, when conducted as stated, will fully dislodge the adherent cells into single cell suspensions, but only break up the spheres into several smaller spheres.

8. As cell counting will be difficult (due to spheres), as a rule, passage half of amount of cells into a new tissue culture flask and top the remaining half of the volume with the conditioned media collected earlier.
9. The remaining half can be used for experimentation or passaged into another flask to bulk up numbers.

3.5 Neuronal Differentiation of AhNPCs

When AhNPC cultures collected between passages 2 to 5 are cultured under differentiating conditions (cell cultured on adherent extracellular matrix in medium supplemented with neurotrophic factors, NGF, NT-3, and BDNF), the cells/spheres reliably differentiate into functionally active neurons and astrocytes.

Two methods of differentiation exist: the whole-neurosphere approach and the single cell suspension approach. In our hands, the whole-neurospheres approach gave significantly higher yields; therefore, this approach will be demonstrated in this section.

3.5.1 Matrigel Plate Coating

Prior to passaging or harvesting the cells for differentiation, coat the culture surfaces with Matrigel.

1. Slowly thaw Matrigel from −20 °C on ice or in the fridge.
2. In cold DMEM:F12 base medium, dilute the Matrigel at 1:40 dilution.
3. Keep the mixture on ice to prevent gel formation.
4. Plate appropriate volumes (Table 6) of Matrigel and leave in a tissue culture hood at room temperature for at least 1 h.
5. Only aspirate the Matrigel prior to plating the cells/spheres.

3.5.2 Cell/Sphere Harvesting and Differentiation Procedure

1. If the spheres are obtained during passaging (Subheading 3.4), start at **step 4**.
2. When harvesting full spheres for differentiation, gently tap the flask and remove all the media and place in a separate falcon tube.

Table 6
Summary of the surface area of the culture wells, as well as matrigel volume and neurosphere/media mixture volumes that are required for neuronal differentiation experiments of AhNPCs

Plate	96-well plate	24-well plate (electrophysiology)	6 Well plate
Surface area (cm^2)	0.33	1.9	9.6
Matrigel volume (µL)	50	300	1,500
Sphere mixture volume (µL)	100	600	3,000

3. Wash the flask once in half the volume of proliferation medium and add to the falcon tube containing the spheres.
4. Gently centrifuge at 30×*g* for 5 min and remove the supernatant.
5. Gently resuspend the spheres in designated volume of neuronal differentiation medium. This volume is dependent on the density estimation of the spheres (*see* **Note 4**).
6. Add appropriate volumes of the above differentiation medium (with the spheres) into designated culture plates (Table 6).
7. Every second day, carefully half change the differentiation medium by taking media from the top of the cultures, so not to disturb the differentiating cells adherent at the bottom of the wells/flasks.
8. Sprouting of neurites are usually visible within a couple of days and migration of cells from spheres are visible within a week.

Note: We have been able to electrophysiologically record from cells differentiated and cultured for as long as 2 months.

3.6 Differentiation Characterization

3.6.1 Immunocytochemistry

To assess the expression of lineage-specific markers/proteins expressed by the cells, cultures are generally processed for immunocytochemistry (ICC). To determine lineage of the cells, we typically use double or triple fluorescent labelling techniques to co-localize marker proteins in the same cell. These stains can be imaged using either a confocal microscope for high quality images or a conventional wide-field fluorescent microscope on monolayer cultures for the determination of co-localization. We recommend using the Alexa Fluor® 350, 488 or 594 conjugated secondary antibodies for visualization and Hoechst for nuclear staining (*see* **Note 5**).

1. At designated time points, aspirate the culture medium and fix the cells in 4 % (w/v) paraformaldehyde (PFA) for 15 min at room temperature.
   (a) To minimize the disturbance to the culture caused by aspiration, we recommend adding equal volumes of 8 % (w/v) PFA straight into the culture medium.
2. Aspirate the PFA and wash three times in PBS with 0.1 % (v/v) of Triton X-100 (PSB-T).
3. Dissolve the primary antibodies specified in Table 7 in PBS-T with 1 % (v/v) goat serum (immunobuffer) and incubate them with the cultures overnight at 4 °C.
4. Next day, wash the cultures three times in PBS-T before adding Alexa Fluor® conjugated anti-mouse or anti-rabbit secondary dissolved 1:500 in immunobuffer.
5. Incubate for 3 h at room temperature or overnight at 4 °C.

Table 7
List of antibodies used for lineage determination in AhNPC cultures

Antibody name	Species	Supplier	Dilution	Lineage
βIII-tubulin	Mouse	Sigma	1:1,000	Immature neuron
MAP2ab	Mouse	Sigma	1:500	Immature-mature neuron
GFAP	Rabbit	Dako	1:2,000	Astrocytes or NSC/NPCs
Nestin	Mouse	Millipore	1:500	NPCs and neuroblasts
Fibroblast	Mouse	Dako	1:1,000	Collagen synthesizing cells

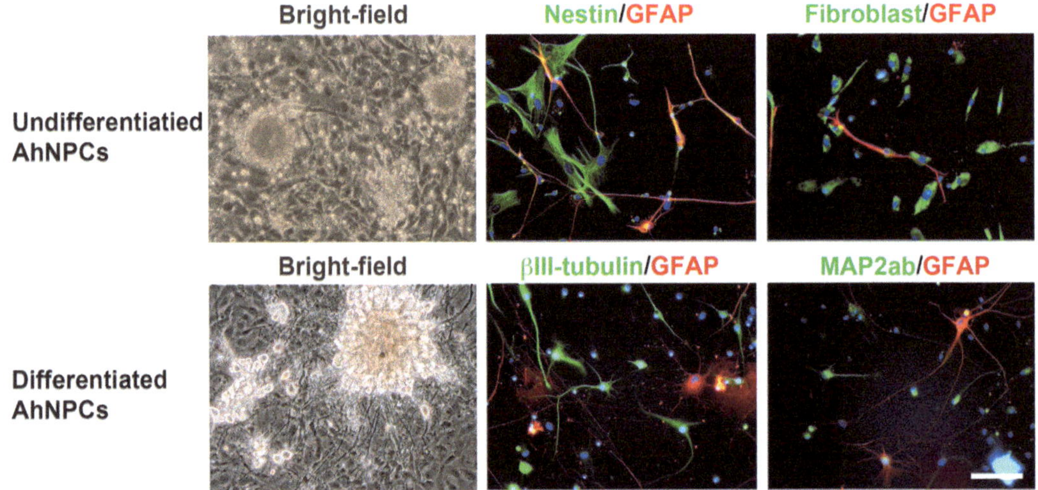

Fig. 5 Photomicrographs of undifferentiated and differentiated AhNPCs. *Top row* represents images of undifferentiated AhNPCs in culture. A bright-field image shows spheres afloat on top of a monolayer of cells. Triple label images show co-localization of Nestin with GFAP in these cultures. Fibroblast marker (Prolyl-4-hydroxylase) is expressed in many of the underlying cells. *Bottom row* represents images from differentiating AhNPCs. Bright-field image clearly shows neurites extensions from an adherent sphere and triple labelled images show βIII-tubulin and GFAP as well as MAP2ab and GFAP staining. Many of the cells do not co-localize in these images, hence representing clear lineage segregation into neurons and astrocytes. Scale = 100 μm

6. Following this incubation, wash the cultures once in PBS-T before adding Hoechst dissolved 1:500 in Tris-buffered saline (TBS). Incubate for 1 h at room temperature and wash 3 times in PBS-T.
7. Now the cells are ready to be imaged. After the final PBS-T wash, store the cells in PBS-T with 0.1 % (v/v) thiomersal in the dark and image using a fluorescent microscope (Fig. 5).

3.6.2 Electrophysiological Characterization

In conjunction with ICC studies, functional analysis using electrophysiological techniques can provide an unequivocal confirmation of the extent of neuronal differentiation. Therefore, where possible, electrophysiological confirmation should accompany molecular studies.

To setup cultures for electrophysiological experiments, we plate cells onto 12 mm round coverslips. This can be achieved by coating the coverslips in a 24-well plate. These coverslips can later be transferred to an electrophysiology rig and the cells can be targeted under differential interference contrast (DIC) microscopy for patch-clamping. Once cells are patch-clamped, a number of functional studies can be conducted under voltage-clamp and current-clamp configurations. For this chapter, we will only concentrate on setups that test for the typical physiological characteristics of a neuron, such as, resting membrane potentials of −50 to −65 mV, action potentials, and fast-inactivating inward and slow-inactivating outward currents.

Cell Culture and Rig Setup

1. AhNPCs are cultured as stated in Subheading 3.5 on glass coverslips coated with Matrigel®.
2. On the day of the recording, setup the rig for electrophysiological recordings.
   (a) Prepare the external solution and continually bubble carbogen into the ACSF.
      - External solutions can be made as 10× stocks (without glucose).
      - On the day of the experiment, the stock can be diluted and glucose can be added to the solution.
   (b) Adjust the flow rate of the external solution to be approximately 2–3 mL/min.
   (c) Prepare the internal solution (usually made up, aliquoted into 1 mL solutions and stored at −80 °C).
      - Optional: To identify or double label patch-recorded cells, an Alexa Fluor® dye can be diluted (1:250) into the internal solution.
   (d) Fabricate the patch-pipettes to yield a pipette access resistance of approximately 6–7 MΩ when filled with internal solution and placed in the external solution.

Patch-Clamping Differentiated AhNPCs

1. Once the rig is ready, remove the coverslips (with the differentiated cells) from the culture well using sterile forceps and placed it onto a recording chamber equipped with DIC optics.
2. Under 40× optics, locate a cell with neuronal-like morphology (numerous neurite outgrowths and axonal-like structure if visible).
   (a) Generally, there are more astrocytes compared to neurons and since they both have similar morphologies *in vitro*, the majority of your recording will end up being astrocytes.

- To increase your chances of patching a neuron, try to look for cells with a distinct axonal morphology.

(b) One can transfect the cells with a reporter gene under the control of a neuronal promoter to increase the chances of recording from a neuron.

3. Once the electrode is in the bath solution, apply positive pressure (e.g., using a syringe) and measure the electrode's access resistance by applying 10 mV steps at 100 Hz. Ideally, you want this value to be between 5 and 10 MΩ.
4. Lower both the objective and the electrode, synchronizing the move so that the objective's focus is always just ahead of the tip of the electrode (*see* **Note 6**).
5. Once near the bottom (i.e., cells), focus onto the cell of interest and slowly position the electrode over the cells.
6. As the electrode approaches the cell, a characteristic dimple will appear on the surface of the cell (due to the positive pressure applied by the tip of the electrode pushing on the cell membrane).
7. At this moment, release the positive pressure.
8. In most occasions, the cell membrane will bounce back and form a seal with the electrode. As the access resistance increases due to the seal, apply a −60 mV holding current.
9. Once the electrode's access resistance reaches 1 GΩ (known as a "gigaseal"), apply short bursts of negative pressure (usually by short suctions), to break the membrane and establish a patch-clamp connection. Please *see* **Note 7** for some trouble shooting if you are having trouble forming a gigaseal.
10. Once the cell is patch-clamped, record the cell's parameters by using a membrane test function (e.g., membrane test algorithm in pClamp8 software (Axon Instruments)).
11. Hereafter, the end user can decide which parameters to test. Some examples for testing neurophysiological characteristics could be:
    (a) Depolarizing voltage-steps to record active current responses such as fast-inactivating inward currents.
    (b) Depolarizing current-steps to record the resting membrane potential and test for the presence of evoked action potentials.
    (c) Test for channel specific activity by using pharmacological blockers of channels (e.g., tetrodotoxin to block voltage-gated sodium channels, hence block action potentials).
12. After the recordings, the coverslip can be place in 4 % PFA and fixed and counterstained using standard staining procedures. This will allow post-patch immunocytochemical analysis.

4 Notes

1. Ethical note for human brain collection.
   Human brain tissue is obtained from the Neurological Foundation of New Zealand Human Brain Bank (Centre for Brain Research, University of Auckland). The University of Auckland Human Participants Ethics Committee has approved the consent and research protocols used in these studies. Our procedures for obtaining brain tissue both from post mortem cases and from surgical specimens are reviewed every 3 years by the regional or University of Auckland Human ethics committee. If the donor is donating surgically resected tissue then informed consent must be obtained before tissue is received. For whole brain donation, the next-of-kin must agree to the donation of brain tissue and must sign a consent form before the brain is removed.
2. Antigen retrieval.
   Antigen retrieval is most often required to detect nuclear proteins. Additionally, we often receive brains that have had a long fixation period and these often require antigen retrieval depending on the epitope of interest. Our experience has been that cell cycle proteins such as proliferating cell nuclear antigens, other cyclins, histones and transcription factor antibodies usually require antigen retrieval to be performed on the tissue before the antibodies will detect the protein of interest. The commonest antigen retrieval we use is citric acid treatment and pH 4.5. Varying both the antigen retrieval solution and the pH is an important first step before beginning large studies that consume lots of tissues and chemicals. Although only one method of antigen retrieval is mentioned above, there are numerous other methods of antigen retrieval that work successfully on different antigens. In addition, for detecting Bromodeoxyuridine (BrdU) tissue must undergo a standard hydrochloric acid pretreatment to enable BrdU presentation, prior to incubation in rat anti-BrdU antibodies. The end user must determine, through using controls, whether their antigen retrieval method is suitable for the antigen of interest. Our experience has been that the above protocol for antigen retrieval is simple, fast and yields the best results on average and makes a good starting point for revealing a wide range of antigens.
3. Immunoelectron microscopy.
   For immunoelectron microscopy, the sections are processed in the same way as the sections for light microscopy up to the stage where they have been treated with the DAB chromogen except that no Triton X is used in the procedure.

4. Sphere-density estimation.
   As whole sphere numbers are difficult to determine, we estimate our plating density based on the number of spheres present in a 10× magnification field-of-view in the original culture. The reader should optimize their dilutions based on the statement below.

   In our experience, 5–7 spheres (of size 100 μm or above) are present in a field-of-view and the pellet was resuspended in equal volume of differentiation medium for plating, 100 μL of sphere/medium mixture will result in approximately 10 spheres per 96-well plates (surface area 0.33 cm^2).

   Therefore, if an average of 10 spheres are present in a single field-of-view, one should resuspend in twice the volume of differentiation media to obtain approximately 10 spheres per 96-well plate.
5. Fluorescent labelling technical note.
   One can design an experiment where Alexa 350 is used in conjunction with Hoechst, but should try to avoid labelling a nuclear antigen, as they fluoresce at similar wavelengths and can be difficult to tease apart when completely overlapped.
6. Electrophysiology.
   Place the cell of interest in the middle of the field-of-view and lift the objective out of the recording chamber and under voltage-clamp configuration, position the recording electrode in the area of interest. One can temporarily increase the intensity of the light to find the region to place the electrode. Lower the objective to find the tip of the electrode.
7. Forming a gigaseal.
   If you are not achieving a gigaseal within 1–2 min of releasing the positive pressure, try patching the cell again by slightly retracting the electrode, apply positive pressure, and repositioning the electrode over the cell again, but this time, closer to the cell. If you are still having trouble, change the electrode. If this still does not work, this could be due to a compromised cell's membrane integrity/health. We suggest you change coverslips and investigate certain elements during your differentiation process that could have compromised your cell's health. Note, this has seldom happened in our hands.

Acknowledgements

The authors wish to acknowledge their gratitude to the Health Research Council of New Zealand for Programme-Grant funding. The Neurological Foundation of New Zealand for funding the Brain Bank. The Hugh Green Charitable Trust for funding the Bio Bank and the Lynette Sullivan Trust for salaries and ongoing support of work on brain diseases.

References

1. Globus JH, Kuhlenbeck H (1944) The subependymal cell plate (Matrix) and its relationship to brain tumors of the ependymal type. J Neuropathol Exp Neurol 3:1–35
2. Eriksson PS et al (1998) Neurogenesis in the adult human hippocampus. Nat Med 4: 1313–1317
3. Reynolds BA, Weiss S (1992) Generation of neurons and astrocytes from isolated cells of the adult mammalian central nervous system. Science (New York, NY) 255:1707–1710
4. Louis SA et al (2008) Enumeration of neural stem and progenitor cells in the neural colony-forming cell assay. Stem Cells (Dayton, OH) 26:988–996
5. Reynolds BA, Rietze RL (2005) Neural stem cells and neurospheres-re-evaluating the relationship. Nat Methods 2:333–336
6. Roy NS et al (2000) Promoter-targeted selection and isolation of neural progenitor cells from the adult human ventricular zone. J Neurosci Res 59:321–331
7. Palmer TD et al (2001) Cell culture. Progenitor cells from human brain after death. Nature 411:42–43
8. Westerlund U et al (2003) Stem cells from the adult human brain develop into functional neurons in culture. Exp Cell Res 289:378–383
9. Moe MC et al (2005) Multipotent progenitor cells from the adult human brain: neurophysiological differentiation to mature neurons. Brain 128:2189–2199
10. Richardson RM et al (2006) Isolation of neuronal progenitor cells from the adult human neocortex. Acta Neurochir 148:773–777
11. Leonard BW et al (2009) Subventricular zone neural progenitors from rapid brain autopsies of elderly subjects with and without neurodegenerative disease. J Comp Neurol 515:269–294
12. Waldvogel HJ et al (2006) Immunohistochemical staining of post-mortem adult human brain sections. Nat Protoc 1:2719–2732
13. Fritschy JM et al (1998) Synapse-specific localization of NMDA and GABA(A) receptor subunits revealed by antigen-retrieval immunohistochemistry. J Comp Neurol 390:194–210
14. Curtis MA et al (2003) Increased cell proliferation and neurogenesis in the adult human Huntington's disease brain. Proc Nat Acad Sci 100:9023–9027
15. Waldvogel HJ et al (1999) Regional and cellular localisation of GABA(A) receptor subunits in the human basal ganglia: an autoradiographic and immunohistochemical study. J Comp Neurol 415:313–340
16. Waldvogel HJ et al (2004) Comparative cellular distribution of GABAA and GABAB receptors in the human basal ganglia: immunohistochemical colocalization of the alpha 1 subunit of the GABAA receptor, and the GABABR1 and GABABR2 receptor subunits. J Comp Neurol 470:339–356
17. Baer K et al (2003) Association of gephyrin and glycine receptors in the human brainstem and spinal cord: an immunohistochemical analysis. Neuroscience 122:773–784
18. Adams JC (1981) Heavy metal intensification of DAB-based HRP reaction product. J Histochem Cytochem 29:775
19. Faull RLM, Villiger JW (1986) Heterogeneous distribution of benzodiazepine receptors in the human striatum: a quantitative autoradiographic study comparing the pattern of receptor labelling with the distribution of acetylcholinesterase staining. Brain Res 381:153–158
20. Faull RLM, Villiger JW, Holford NHG (1987) Benzodiazepine receptors in the human cerebellar cotex: a quanitative autoradiographic and pharmacological study demonstrating the predominance of type 1 receptors. Brain Res 411:379–385
21. Faull RLM, Villiger JW (1988) Multiple benzodiazepine receptors in the human basal ganglia: a detailed pharmacological and anatomical study. Neuroscience 24:433–451
22. Faull RL, Dragunow M, Villiger JW (1989) The distribution of neurotensin receptors and acetylcholinesterase in the human caudate nucleus: evidence for the existence of a third neurochemical compartment. Brain Res 488:381–386
23. Faull RL et al (1993) The distribution of GABAA-benzodiazepine receptors in the basal ganglia in Huntington's disease and in the quinolinic acid-lesioned rat. Prog Brain Res 99:105–123
24. Glass M, Dragunow M, Faull RLM (2000) The pattern of neurodegeneration in Huntington's disease: a comparative study of cannabinoid, dopamine, adenosine and GABA(A) receptor alterations in the human basal ganglia in Huntington's disease. Neuroscience 97: 505–519
25. Waldvogel HJ et al (1990) GABA, GABA receptors and benzodiazepine receptors in the human spinal cord: an autoradiographic and immunohistochemical study at the light and electron microscopic levels. Neuroscience 39:361–385
26. Gibbons HM et al (2007) Cellular composition of human glial cultures from adult biopsy brain tissue. J Neurosci Methods 166:89–98

Part VI

Clinical Application of Neural Precursor Cell Therapy

Chapter 18

Clinical Trials for the Treatment of Spinal Cord Injury: Not So Simple

Alan Mackay-Sim and François Féron

Abstract

The fast pace of research in stem cell biology and regenerative medicine is feeding hopes of the scientific community and the public that a new revolution in treatments is upon us. There are increasing numbers of examples of stem cell therapies that are effective in treating animal injuries and diseases. There is an expectation that stem cell transplantation will soon be commonplace in the human clinic, especially with the beginnings of clinical trials of embryonic stem cell transplantation for bone repair, spinal cord injury, macular degeneration, Stargardt's disease, and Batten's disease. This may be an appropriate point at which to review our experiences in moving from the lab to the clinic to initiate a Phase I clinical trial of autologous olfactory ensheathing cells in spinal cord injured humans.

Key words Spinal cord injury, Olfactory ensheathing cell, Adult stem cell, Embryonic stem cell, Induced pluripotent stem cell, Clinical trial

1 Introduction

With the rapid advances in stem cell biology and many demonstrations of efficacy of stem cell transplantation in animal models, there is much hope for the quick translation of stem cell therapies into the clinic. We recently completed a first-in-man trial of autologous olfactory ensheathing cell transplantation into the injured spinal cord in human paraplegia [1, 2]. While not strictly "stem cells," many of the same practical issues arise when moving from animal models to human clinical trials. One is easily excited by the potential of cell therapies in animal models, but there are many practical and philosophical issues that must be surmounted before engaging in a clinical trial. The processes required of human clinical trials should provide a "reality check" when lab researchers declare that a particular stem cell therapy is "5–10 years" away from clinical application. Perhaps in almost all cases the lead-time for clinical application when discussing the latest "breakthrough"

Brent A. Reynolds and Loic P. Deleyrolle (eds.), *Neural Progenitor Cells: Methods and Protocols*, Methods in Molecular Biology, vol. 1059, DOI 10.1007/978-1-62703-574-3_18,

in an animal model the reality is going to be closer to 20 years "if all goes well" and "funding is found for the clinical trials." We discuss here some of the issues we faced in a trial of cell transplantation into the spinal cord.

2 Who Makes the Decision to Start a Clinical Trial?

Stem cell researchers used to the model of a single individual driving a research program will need to accommodate a complex, collaborative decision about applying their ideas in the clinic. The decision to conduct a clinical trial is soon out of the control of one individual and becomes a collective decision of a large number of participants, researchers, and regulators. As in all research it also involves a decision of agencies to fund the expenses of the trial.

At the start of moving out of the lab and into the clinic the researcher must build interest and agreement with a team of clinicians. In our spinal cord injury trial the research team included spinal injuries physicians, neurosurgeon, spinal orthopaedic surgeon, otolaryngologist, trials manager, as well as the authors, lab-based neuroscientists. Additionally, a team of clinical assessors was required including neurologist, neurophysiologist, physiotherapist, occupational therapist, social worker, and psychiatrist. Regulatory requirements and prudence required an additional spinal injuries physician to independently oversee the health and well-being of the trial participants.

Animal researchers will already be familiar with the necessity to comply with and get agreement from institutional ethics committees and bio-safety committees, so they are already familiar with regulatory factors that affect their decisions about research. Clinical trials require further regulation at the state/national level through such agencies as the FDA and its equivalent in most countries that regulates medicines and clinical products. Advancing a clinical trial requires negotiation and review of the study protocol at multiple levels. It is also highly recommended to assign the responsibility of monitoring safety and efficiency outcomes to an external independent group and preregister the main elements of the research protocol in a public dedicated database (e.g., www.clinicaltrials.gov). In our case this involved university and hospital ethics committees, external reviewers commissioned by the ethics committees, and oversight by the Therapeutics Goods Administration of Australia.

Most importantly, a clinical trial requires the fully informed consent of the participants and, in the case of spinal injured patients, this may be a decision of families and carers who help patients organize their lives to accommodate because the demands of trial participation.

3 Scientific Evidence

When is their "enough evidence" to envisage a clinical trial of a new cell therapy? What "kind" of evidence is required? Are the animal models appropriate? Do we need to go to primates or another animal with a close-to-human physiology?

The usual expectation in drug trials is that there should be robust evidence for safety and efficacy in animal models and that there should be some understanding of mechanism. Certainly safety is paramount. Efficacy in animal models is expected, although the list is very long of treatments that fail at the Phase II and III (efficacy) stages in human clinical trials. Similarly there is a long list of drugs in uses for which the knowledge of "mechanism" of action in humans is poorly understood because they have actions not originally contemplated.

When considering a cell treatment for spinal cord injury, what "standard" of evidence of efficacy do we require? Do we require that a treatment show full recovery of locomotor activity—coordinated walking on the back legs, fine motor control of the fore paw? Perhaps we expect full control of breathing and autonomic function in cervical spinal cord injuries? Do we require evidence that the treatment fully reconstructed the cortico-spinal motor tracts or the spinothalamic sensory tracts? Do we also need evidence that all midbrain and brainstem motor and sensory tracts are restored?

When considering an animal model for human spinal cord injury, what is the "appropriate" animal model? Is it enough to see efficacy in mouse or rat models? Are they "too small"? Do we need evidence in a larger animal? Do we need evidence in a primate or a pig? What is an "appropriate" injury? Regeneration and regrowth of axons across a transected spinal cord proves the capability of the treatment to assist axon regeneration, but human injuries rarely involve a surgical transection. Perhaps a model can illustrate "mechanism" but is the same mechanism seen in a different species and different type of injury? Contusion and compression models more realistically model some aspects of human spinal cord damage but again, they do not model all the attendant damage caused when a human chest meets an immovable object at 50 km per hour and the delay between the accident and the treatment in the intensive care unit. Furthermore, there is a phylogenetic decline of regenerative capacity, leaving paralysed humans with a limited to recover. In an animal model, what is an equivalent delay between injury and treatment in an "acute treatment" model? In a "chronic treatment" model, what is the animal equivalent period to a human. What "clock" do we use? The ratio of life-span of animal model and human? The ratio of metabolic rate?

A conservative, risk-averse decision may be to delay translation until we "understand everything" about the injury model and the cell treatment in an animal model and further, we fully understand how the human injury and response to injury differs from the animal model. Realistically, neither of these conditions will ever be met. The question then arises as to how much uncertainty is acceptable and what evidence is it practical to obtain for efficacy, to be balanced against evidence for the risk of harm and the potential benefit gained, in patient health and in knowledge.

4 Patient Selection

Which and how many patients do you select? In human spinal cord injury this means, what kinds of injuries and how long since injured? This decision depends partly on whether you are undertaking a Phase I (safety) trial or a Phase II (efficacy) trial. In Phase I trials safety and risk to the patient are paramount. As cell transplantation is an invasive procedure you need to select patients for whom the risk of adverse consequences is minimized. The general risks associated with anesthesia and surgery cannot be minimized specifically, but the specific risks must be managed in the clinical trial. In spinal cord injury, thoracic injuries are considered to be the optimal choice to minimize the risk of functional impairment [3] because transplant-induced damage above the level of injury might be expected to have minimal effects only on intercostal muscle functions. In contrast, transplantation into cervical injuries risks major hand or arm function and sensation, loss of breathing and autonomic functions. In Phase II trials, in which efficacy is being tested, thoracic injuries are less suitable because thoracic injuries are usually very large because of the forces required to dislodge the spine in the chest. The resultant damage renders the repair task much larger both for endogenous repair and for any treatment, including cell transplantation. A problem then arises because the risk of transplantation-induced damage increases in cervical injuries, so a follow-up from a Phase I trial in thoracic injuries might be a Phase I trial in cervical injuries, or a Phase I/II, rather than simply a Phase II. One way to handle this is a staggered entry trial in which safety can be monitored closely in the early recruits and recruitment to the trial stopped if any safety issues arise. Alternatively, if the procedure has been successfully assessed in an animal model, invasiveness might be dramatically reduced if cells can be transplanted within the cerebrospinal fluid, via a spinal transdural delivery.

Another aspect of patient selection in spinal cord injury is the clinical severity of injury. Spinal cord injuries are considered as "complete" and "incomplete," depending on the loss of motor and sensory function below the level of injury. The classification is based on clinical signs and symptoms, rather than being based on physical connection of the spinal cord. Physical connection is still hard to ascertain from MR imaging. What severity of injury should be chosen for clinical trials, based on a scientific understanding from animal models? Animal models can be incomplete (e.g., hemi-section, and moderate contusion) or complete (full transection severe contusion). In most studies the best recovery in animal models is after incomplete injuries. It makes sense then to undertake clinical trials in incomplete injuries in humans as they might be expected to have the most to gain. They also have the most to lose should the experimental treatment go wrong. Again there is a trade-off between potential benefit and potential risk in patient selection.

How long after an injury should one intervene with an experimental treatment? One guide may be the evidence from the animal model. If the treatment works acutely in an animal then an acute intervention may be justified in humans. Then considerations about what "acute" means in an animal model and what it means practically in the human case. How much leeway is there? This question may be addressed partly by understanding the potential mechanism of action of the treatment such that a time-window can be established. What about a treatment for "chronic" injuries? Is there animal evidence for treating a human 1, 6, 12 months, years after injury? When would cell transplantation that is effective in a rat 1 month after injury be effective in a human? If a rat lives 3 years and a human lives 90 years does 1 month of rat life translate into 30 months of human life? Or does the rate of inflammation and repair in the injured spinal cord operate on a different clock speed?

There are ethical and practical considerations that also affect patient selection and the timing of treatment after injury. In most cases there is a slow, often small, unpredictable improvement in clinical signs and functions following the initial diagnosis at the time of injury. Depending on the severity of the injury this improvement has a different time-course, with more severe injuries having less improvement and finding a stable plateau sooner. The cause of improvements is not known, but it suggests physiological changes can occur for up to a year after spinal cord injury. It is possible that a cell transplantation therapy may be more effective when transplanted during this period of spinal cord change and readjustment. On the contrary side, cell transplantation during this period may interfere for the worse with normal physiological changes, leading to adverse outcomes.

The natural history of recovery varies according to the type of injury (complete vs. incomplete) and the timing of grafting (acute vs. chronic). The size of the cohort should be estimated accordingly. Any therapy performed during the acute phase in patients with incomplete functional impairment (ASIA B or ASIA C) will require a very large number of subjects [4]. Conversely, statistically significant outcomes may be detected in relatively small populations of patients with a complete injury (ASIA A) at the chronic phase (over 6 months). Thus, before planning a trial, it is crucial to get advice from an experienced biostatistician. In addition, since the annual incidence of spinal cord injury is low, the investigators may have to opt for a coordinated multicenter trial. As a consequence, in every site, cell culture specialists should be thoroughly trained in advance and center-to-center variations should be carefully and regularly scrutinised.

Soon after spinal cord injury is the period during which an injured individual is most emotionally vulnerable and having to adjust to the huge changes in their life—physically, emotionally, socially. Compared to patients with older injuries, early patients may not be able to give rationally based informed consent. They and their families may also be less inclined to give consent for an experimental trial treatment until they understand and know the outcomes of the recent injury. Practically this limits the pool of patients to enrol in a clinical trial. Patients with older injuries are better able to give informed consent because the consequences of the injury are more personally understood. This raises the available numbers of volunteers to choose from. Balanced against this may be a potential loss of spinal cord plasticity in older injuries, such that the chances of success may diminish with time after injury. Another factor is the baseline stability of older injuries that makes it statistically easier to measure a mean difference between groups when the baseline variability is less (as in older patients) compared to early injuries in which patient signs improve with time. As a consequence of these different forces, selection of patients is a complex balancing act between changing baseline clinical scores, time-dependent physiological changes, time-dependent changes in capacity to consent, time-dependent available pool of patient volunteers, the relative risks associated with each of these factors, and the relative benefits of intervention, set against an incomplete background of data generated in imperfect animal models (Fig. 1).

5 Autologous or Allogenic Cells?

The researcher wishing to take their stem cell candidate into human clinical trials will be driven by their own research agenda that may already be directed at either autologous or allogenic cell types. It is worth considering the strengths and weaknesses of both approaches in the human clinical context.

Autologous stem cell transplantation is limited to "adult" or "tissue-derived" stem cells. These may be derived from bone-marrow, adipose tissue, skin biopsy, or biopsy of the olfactory mucosa. The main advantage of autologous cells is that they are immune-matched. The main disadvantage is that they are not available at the time of diagnosis, ready for immediate transplantation. Allogenic cells have the exact reverse advantages and disadvantages,

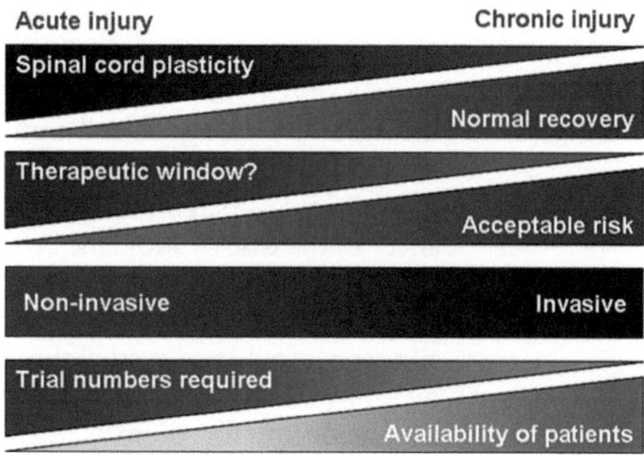

Fig. 1 When to intervene? After spinal cord injury there will be many interrelated factors that will affect the timing of any cell transplantation therapy. One axis might be decreasing plasticity of the spinal cord as it undergoes the normal course of recovery. A second axis might be a reducing therapeutic window as the injury becomes more chronic, faced against an increasing acceptable risk profile based on ethical and social factors. These considerations might suggest that noninvasive procedures are more practical and therapeutic soon after injury, whereas invasive procedures become more practical and acceptable later. When these decisions are made other practical factors come into play. Because of the variability in the natural course of recovery, the numbers of patients required to show a statistical effect may be large initially and decrease with time and recovery. Against this is the availability of patients which may vary because there are fewer acute patients than chronic, and acute patients may be less willing to volunteer

being immediately available but potentially immunogenic. Immunogenicity of allogenic transplantations could potentially be minimized by developing banks of cells that could be immunotyped and, with a large enough bank, most individuals could be matched to a banked cell line. Another solution is to immune-suppress the recipient but this raises ethical and practical difficulties for the patient because their immune system is compromised. This is of particular concern in spinal cord injury where patients are at very increased, even mortal, risk of infections from catheters, pressure sores, and lung immobility.

There are regulatory differences in the application of autologous and allogenic stem cell sources. In most jurisdictions stem cell therapies will be judged to be "products" and regulated as such. Cells that are removed from the body, grown in vitro and reentered into the body in a region they are not normally found are such "products" and are subject to regulation under "Good Manufacturing Practice." These regulations normally require cells to be "manufactured" under closely defined conditions within an approved facility which undergoes standardized testing of quality control and quality assurance at all levels of management through to daily activities of cell culture. Each cell "batch" must be tested and certified not only for microbial impurities and safety but also for "purity" based on a set of defined criteria. These regulations impose large costs on each batch of manufactured cell products. In the case of autologous cell transplantation, each patient requires his own "product" and so each transplantation surgery acquires the full cost of batch manufacturing and testing and quality assurance. In the case of allogenic cell transplantation, the costs of batch manufacture can be amortized across as many patients into whom the cells can be transplanted. Commercial models therefore favour allogenic, "cell-in-a-bottle" therapies that can be produced in large quantities, batch-checked once and distributed to many patients. The trade-off for them is managing potential immunogenicity. This model requires a stem cell source that can be produced in large volumes, such as embryonic stem cells and iPS induced pluripotent stem cells. Clinical translation of other allogenic stem cells, such as mesenchymal stem cells, may be limited by the inability to grow enough cells to amortize the cost of batch testing across enough patients. Autologous stem cell therapies may be limited in both dimensions—not available immediately and requiring onerous batch-testing and the costs involved.

Ideally, a prospective clinical trial should be randomized, double blinded, and placebo-controlled [5]. Such a procedure is relatively straightforward when the therapy is based on the oral administration of a drug with a safe adverse event profile. It is, however, a complicated issue when the trial includes an open surgical manipulation. Inclusion of a sham surgery control group may subject patients to various risks such as infections, pneumonia, autonomic dysreflexia and other postoperative complications. On the other hand, it is established that a simple surgery, independent of any injected cell, may alleviate spinal compression, eliminate unwanted tissue adhesion, enhance the flow of cerebrospinal fluid, enhance the mood of the patient and, as a result, improve the outcome. As a general rule, US investigators tend to include a sham surgery group while, in other countries, clinicians are less predisposed to do so. Our trial comprised a control no-treatment group, without sham surgery, that matched to the experimental group for patient demographics and injury characteristics. This group allowed the assessors to perform single-blinded and reliable measures of research end points. Trials that do not integrate control patients and blinded assessments will yield results of poor value and no valid conclusion on the potential benefit of the therapy.

6 Who Pays for Stem Cell Clinical Trials?

Typically the costs of clinical trials are paid by patent holders who expect to benefit from the product, once proven in clinical trials. The patent holders in stem cell biology are potentially widely distributed, with patents covering cells, methods of cell generation and methods of transplantation. The commercial model requires that a company has patents or patent licences that convince shareholders or venture capitalists that their investments will be rewarded by exclusive profits. This makes the "cell-in-a-bottle" model favourable because a ready-made, batch-tested, allogenic cell is available as soon as it is needed, preferably for more than one indication. For many cell therapies, it is hard to envisage that this commercial model will work, certainly not until the immunogenicity of allogenic cells is overcome. For many stem cell applications, autologous cells may be required. Because of the high cost of cell manufacture these treatments will be costly and, even if effective, the costs may prevent their widespread clinical application. On the other hand, if a treatment was effective enough to reduce the future costs of patient care there may be incentive for large patient care organisations (commercial or publically funded) to pay the high costs now to allay continuing costs for patient care over a life-time. These issues may seem remote from the stem cell researcher doing animal studies but those interested in moving "from the lab to the clinic," knowledge of the complexities of delivering in the clinic may guide the nature of research in the lab.

References

1. Feron F et al (2005) Autologous olfactory ensheathing cell transplantation in human spinal cord injury. Brain 128:2951–2960
2. Mackay-Sim A et al (2008) Autologous olfactory ensheathing cell transplantation in human paraplegia: a 3-year clinical trial. Brain 131:2376–2386
3. Tuszynski MH et al (2007) Guidelines for the conduct of clinical trials for spinal cord injury as developed by the ICCP Panel: clinical trial inclusion/exclusion criteria and ethics. Spinal Cord 45:222–231
4. Fawcett JW et al (2007) Guidelines for the conduct of clinical trials for spinal cord injury as developed by the ICCP panel: spontaneous recovery after spinal cord injury and statistical power needed for therapeutic clinical trials. Spinal Cord 45:190–205
5. Lammertse D et al (2007) Guidelines for the conduct of clinical trials for spinal cord injury as developed by the ICCP panel: clinical trial design. Spinal Cord 45:232–242

INDEX

Brent A. Reynolds and Loic P. Deleyrolle (eds.), *Neural Progenitor Cells: Methods and Protocols*, Methods in Molecular Biology, vol. 1059, DOI 10.1007/978-1-62703-574-3, © Springer Science+Business Media New York 2013

If you have any concerns about our products,
you can contact us on
ProductSafety@springernature.com

In case Publisher is established outside the EU,
the EU authorized representative is:
Springer Nature Customer Service Center GmbH
Europaplatz 3, 69115 Heidelberg, Germany

Printed by Libri Plureos GmbH
in Hamburg, Germany